The Connexin Connection

Your Health Begins with Molecular Fitness

Darrell L.Tanelian, M.D., Ph.D

GOM MEDPRESS

GOM MEDPRESS
An Imprint of Gom Publishing
PO Box 211110, Columbus, Ohio 43221

Phone: 866.466.2608
Email: communications@gompublishing.com
Internet: www.gompublishing.com

Library of Congress Cataloging-in-Publication Data available upon request.

ISBN 1-932966-50-1

First Gom MedPress printing: February 2006

This book is printed in the United States of America on acid-free paper.

Cover designed by Matt Alford, Alford Designs.

Acknowledgements

For those in my life who have displayed the courage to trust themselves and others, and are always willing to timelessly give of themselves. Thank you: Carole Hoffman, Harry Lynch, Marion Carol, Helen Houp, Howard Reisman, and Galite Reisman.

For those who trusted in themselves and transformed the world yet never received the trust of others. Thank you: Aristotle, Anaxagoras, Hippocrates, Galen, Giordano Bruno, Copernicus, Jacques Cartier, Galileo, Ignaz Semmelweis, Nikola Tesla, Niels Bohr, and Alan Turing.

The Connexin Connection is a transformation of an earlier version of this book written in 1999, also called *The Connexin Connection.* Now in 2005, after five years of testing and additional scientific research has been completed by the scientific community to validate predictions I made regarding the eight major diseases back in 1999, this book is ready for presentation to the world.

"PERHAPS the sentiments contained in the following pages, are not yet sufficiently fashionable to procure them general favor; a long habit of not thinking a thing wrong, gives it a superficial appearance of being right, and raises at first a formidable outcry in deference of custom. But the tumult soon subsides. Time makes more converts than reason."

Thomas Paine
Common Sense, February 14, 1776

Contents

Section II The Connexin Molecular Fitness Lifestyle

Prologue

"The first wealth is health."

Ralph Waldo Emerson

"When health is absent, wisdom cannot reveal itself, art cannot become manifest, strength cannot be exerted, wealth is useless, and reason is powerless."

Herophilus, 300 B.C.

THE CONNEXIN CONNECTION

The Connexin Connection presents a dramatically new concept underlying the maintenance of health, the prevention of disease and the extension of life. This concept is derived from the integration of thousands of research studies by many scientists, including decades of research on the connexin molecule.

First, all of the major diseases of modern man are related to the failure of our cells, the basic building blocks of our bodies, to properly perform all of their normal functions. When our cells communicate with each other and perform their normal functions they are said to be healthy and correspondingly we enjoy the vigor and energy of life. When our cellular function becomes impaired they stop communicating and we develop disease.

Second, the ability of our cells to communicate and perform all of their functions is determined largely by the effective functioning of the molecules that connect our more than one trillion cells together and allow the cells in our body to function as a unified whole. The scientific community has given these molecules the name **connexins.** While not the sole essence of life, they play a major role in our health and life. They play a preeminent role in cellular communication, integration and function.

Third, our diet and other lifestyle choices largely determine the integrity of these interconnections between our cells – the effectiveness of our **connexin** molecules. There is abundant scientific evidence that by implementing certain lifestyle choices we can enhance **connexin** function. This will lead to prevention of disease, maintenance of health and extension of our lives.

Fourth, *The Connexin Connection* presents a program for restoring your body to function as it was originally designed to function at the molecular level by your genes and the biophysics of nature. This program is unique in that it integrates science, medicine, nutrition and technology as never before and explains how many seemingly independent beneficial health platforms are all united through the connexin molecule. Now, health, disease, fitness and wellness will make sense to you. You will learn to measure not only your own vital signs, but also track your doctor obtained laboratory tests and medical records and have secure access to them anywhere in the world. You will learn about and measure real indices of your health such as blood pressure, heart rate, blood glucose, body mass index (BMI) and pH (see chapters 4, 5, 11). All of these measures are reflective of the quality and quantity of your diet and health. This data will provide you with real feedback about your state of health and allow you to optimize your health and wellness through enhancing connexin function.

When you embrace *The Connexin Connection* you will experience a new state of well being: **molecular fitness**, a state of optimal health and wellness, from the inside out. All of the essential components lie within your grasp. The choice is yours!

The Connexin Connection concept rests upon thousands of years of man's endeavor to understand his body coupled with decades of international scientific research into the connexin molecule. I have distilled the essence of thousands of scientific articles into the foundation of this health and wellness program. If you are interested in the underlying research, I invite you to reinforce and advance your knowledge of the connexin molecule by referring to the 300 references cited in this book.

By using HealthConnexin.com, you can link directly to over 5,000 connexin references authored by hundreds of prominent scientists.

By and large, the functioning of your connexin molecules tells the story of your life: past, present and future. You are a molecular product of your family, parents and yourself; and moreover, you will influence your children's molecular composition. *The Connexin Connection* is for those who care enough about themselves, their family and others to take responsibility for their health.

The Connexin Connection can transform you and your family's health and wealth. It can transform the future of our nation and the world by curtailing and reducing the ever spiraling healthcare expenditures which are crippling our economy and are projected to have ever more dire consequences for our children's future. Open your mind and focus your attention and energy on this solution to the improvement of your health and the health of our nation. Health is your single greatest asset and it's our greatest collective asset. Embracing the Connexin Lifestyle can make a substantial improvement in our lives!

"So how and why did I develop this health and wellness program?"

In over 30 years of practicing medicine, training physicians and scientists, and being actively involved in scientific and medical research, I came to believe that I was in a system that benefits from illness, not wellness; sickness, not health. I decided that healthcare needed to move in a different direction and needed to be put on a better course.

The Connexin Connection is my synthesis of these many years of science, medicine and engineering. This book is only the beginning of what I hope to be a major paradigm shift in thinking about healthcare. This shift does not stop at just identifying the underlying basis for health and wellness, or by pointing out the current short comings of our current healthcare system. Rather, through years of delivering healthcare, I have recognized many of its underlying problems and have worked to create a comprehensive program towards their correction. HealthConnexin.com was created to unite monitoring, data storage and data transmission technology with healthcare, nutrition, exercise and stress reduction to enable you to monitor and optimize your health, wellness and performance. Sophisticated, yet simple to use monitoring tools allow you to actually see how your body's internal health is improving as your sense of well being also improves!

My journey began at Stanford University where I completed my education (B.S., M.D., and Ph.D.) and training (internship, residency, and fellowship). Subsequently, I held professorships and faculty positions in Neurology, Neuroscience, Engineering, Anesthesiology and Pain Management at Stanford University School of Medicine and The University of Texas, Southwestern Medical School.

Even with all this science, I was never satisfied with the explanations of life or health that I was taught and I was limited to in my practice of medicine. I have routinely witnessed unnecessary suffering of patients and their families. Emotionally, intellectually and practically I questioned the solutions I had to offer them. It became clear to me, that in order to pursue my dream of conquering illness, I had to step out of the conventional medical paradigm. I needed to obtain the intellectual freedom necessary to develop a solution to our number one challenge of the 21st century – the prevention of illness. I have spent the last 7 years conducting independent research into health and nutrition. The result: the development, the testing and the refinement *The Connexin Connection* program, which I know will improve your health, and may remedy our nation's healthcare crisis.

On the journey to develop *The Connexin Connection*, it became clear to me that most of us in industrialized societies have an inability for our cells to function as an effective, unified whole (impaired connexin molecule function) and that this dysfunction is associated with the eight major diseases of modern man - cancer, heart disease, diabetes, arthritis, osteoporosis,

Alzheimer's, hypothyroidism (obesity), and ocular disease (cataracts).

The lifestyle outlined in *The Connexin Connection* aims to make our cells work as an effective whole. To date many have scientifically, medically, and functionally validated and confirmed the health and performance benefits of embracing *The Connexin Connection* program. I have personally witnessed these benefits, and am myself a recipient of these benefits.

The Connexin Connection Molecular Fitness Lifestyle makes us well by optimizing the performance of our keystone of health and life, the connexin molecule. It is not a miracle that will eliminate every unhealthy moment, nor will it totally reverse any unfavorable choices you may have made regarding your health in the past. However, the earlier in life you begin the connexin lifestyle, the greater it's preventative value. But, it is never too late for you to benefit!

The Connexin Connection provides you a clear-cut road map and guide with which to improve your health and wellness through optimized connexin function. There is much scientific evidence that with excellent management of our health and life we will have open, effective, excellent communication between our cells, with the result being superior health, protection from disease and the potential for the extension of life. But this all greatly depends on the care of your connexin molecules. When your molecules function optimally (we call this *Molecular Fitness*), your health will be enhanced, and disease will be reduced. This program of connexin enhancing foods, supplements, and lifestyle interventions has been carefully developed and tested to provide you an easy to use solution for achieving the quality and quantity of life you deserve. However, just as there is no "free lunch", there is no miracle cure or quick fix. Remaining healthy is a lifelong commitment to yourself and your loved ones. I look forward to your participation in improving the quality of your life and the lives of your loved ones.

"Restoration of the connexin molecule will herald a renaissance in our health, in our healthcare, and in the future of mankind."

Darrell L. Tanelian, M.D., Ph.D.

1 INTRODUCTION

"He who enjoys good health is rich, though he knows it not."

Italian Proverb

"What is now proved was once only imagined."

William Blake

Life as we know it began as a series of transformations some 2 to 3 billion years ago, and will continue long after we have gone. In the last 600 million years, two of the most pivotal life transformations that took place were: **Cellular Transformation**, from a single cell organism to a multi-cellular organism; and **Species Transformation**, a divergence from Neanderthal to Homo Sapien, resulting in today's modern man. **Health Transformation** is what I call our current epidemic of chronic illness, which will have a major impact on future generations unless dramatically improved.

Cellular Transformation began coincident with the mysterious appearance of the connexin molecule. When 6 connexin molecules combined for the first time into a unit called a connexon, the resultant junction between neighboring cells allowed direct cellular communication to take place. More than one cell could join together, with the resultant multicellular organism functioning as a unified whole. Over time, many, millions, even billions of cells merged together to establish the life-forms we know today. The sum of these merged cells, the whole, became functionally greater than the sum of the individual parts. These multicellular life forms (worms, insects, fish, birds, other animals, humans, etc.) are what we recognize as "life". All of these multicellular organisms and virtually all cell types within those organisms, contain connexins.

This transformation from single cells uniting to become man didn't happen over night! In fact, it took about 600 million years for cells to become arranged in an order that could be considered human. Man is thought to have emerged, albeit in a primitive state, only two to three million years ago. And as of 100,000 years ago, only two main species of man remained to roam the earth as competitors; Neanderthal and Homo Sapien. Hence began what I call **Species Transformation**.

The earth was a world of great abundance, filled with a great diversity of plants, animals, and nutrient sources. Population density was low. I believe the most fit competitor destined to become modern man was the species that selected the most complementary diet for their century's old, genetically pre-determined biochemistry and physiology.

Let us compare our two final competitors, Neanderthal and Homo Sapien. Neanderthal being stocky and heavily muscled would be considered overweight by modern standards. Examination of their skeletons reveals a lifespan of 30 to 40 years. **The Neanderthal diet was 90% protein from meat.** Despite having a larger brain than modern man, it did not provide greater intelligence. Neanderthal lacked innovation in that he used primitive tools, and never developed art, architecture, boat building or formal language. In sharp contrast, Homo Sapien lived into his 60's, practiced long-distance trade, and developed art, language architecture and technology. **The Homo Sapien diet consisted of fruits, vegetables, seafood and meat — a balanced Mediterranean diet.**

I believe the diet adopted by Homo Sapien explains the vast differences between these two species of man. Homo Sapien's improved diet was connexin enhancing, a boon to his natural physiology and biochemistry. This led to improved fitness and superiority.

Homo Sapien initially arose in Africa, migrating to the Fertile Crescent and then to the Mediterranean. Later, he populated and left his genetic legacy throughout Europe and Asia. Homo Sapien's original diet was in essence the original Mediterranean diet; which has been shown today, by many medical and scientific studies, to reduce the occurrence of disease and to enhance longevity. This diet also promotes connexin function. I believe this connexin enhancing diet gave Homo Sapien an advantage over Neanderthal, allowing the Homo Sapien species to prevail.

While consuming this connexin enhancing diet, man exhibited some of his greatest advancements, beginning around 40,000 years ago in the regions today known as France and Spain. These technological and cultural advancements set the stage for man's colonization of the rest of the world, and with him he brought his connexin healthy diet.

More recently, "advances" in agriculture and in the domestication of animals and especially the development and growth of processed fast foods over the last 60 years, have taken us away from the connexin healthy diet that provided for the development of our species. Our fast-food world has resulted in nutritional scarcity for many, if not most. **The irony is that with all the knowledge we have attained, and the advent of "health conscious" food markets, we should be living in the greatest era of human nutrition and health ever seen on this Earth.**

This is not the case. Our modern diet has led us astray. Consider what I regard as the eight major diseases of modern man — heart disease, cancer, diabetes, osteoporosis, arthritis,

Alzheimer's, hypothyroidism (obesity) and ocular disease (cataracts). **Each of these diseases can be linked to our modern diet and to a decrease in connexin function.** Epidemiological data shows that the prevalence of all of these illnesses is on the rise and that modern medicine is neither preventing nor curing these conditions. **It is interesting that as we remove connexin enhancing molecules from our diet and the rest of the world follows suit, there is parallel expansion of global illness.** We are now in the midst of a **Health Transformation** that must be reversed.

Take cancer and heart disease, for example. In 1900, cancer was responsible for only 4 percent of all deaths in the United States and heart disease killed 9 percent. By 2005, cancer is expected to contribute to 25 percent of all deaths, with all cardiovascular disease being responsible for 38 percent. Cancer is especially interesting since the mortality rate from the disease has steadily increased since 1900, from a death rate of 64 per 100,000 to 216 per 100,000 in 2001. That's an increase of 225 percent! By 2001, the lifetime risk of being diagnosed with cancer had jumped to 50% or 1 in 2 for men and slightly over 33% or 1 in 3 for women.

What about the cost? Look at the rise in healthcare spending since 1960, when its total cost for everyone in the U.S. was $25.2 billion per year, or 1.1% of the Real Gross Domestic Product. By 2003, the cost rose to $1.67 trillion or 15.3% of the GDP — an almost 15-fold increase and impact on our economy. And it is projected to hit $2.75 trillion by 2010. We all know how much health insurance and prescription drugs cost these days. The per capita healthcare expenditure is expected to be $6,546 this year (2005). Corporate healthcare expenses now exceed corporate profits. The average American now spends more on healthcare than taxes.

A recent Harvard study showed that 50 percent of all personal bankruptcies in the United States were a result of illness and associated healthcare costs in middle class, insured people. Furthermore, 29 million Americans are in "medical debt", paying bills to healthcare providers or incurring large credit card debt or loans against homes related to medical costs. We are mortgaging our children's futures to pay for new drugs, elaborate tests and high-tech operations that wouldn't be needed nearly as often if we would correct our diet and lifestyle.

There have been many proposed revelations or truths regarding health and illness that have surfaced and subsided during my life. However, out of those many revelations five have stood the test of time and continue to resound ever louder:

1) Health is determined primarily by nutrition and lifestyle.
2) "Modern medicine" does little to prevent illness and overall may contribute more to illness than wellness.

3) On an individual basis and collectively as a society we can no longer afford the cost of illness.
4) Health and wellness involve personal responsibility and commitment.
5) The greatest freedom attainable comes from being healthy, having a healthy family, and staying healthy throughout our lives.

This book, *The Connexin Connection*, explains the interrelationship between **life, health, disease and death** and their relationship with the connexin molecule. Combined, these four aspects of life represent **a unified whole,** of both your body and your life. They are not independent stages of life as they have been frequently represented to us. Our management of life and health can have a profound effect on our experience with disease and death.

Disease has been historically presented as an unpredictable, unrelated, artificial and mysterious event, when in fact our failure to manage life and health has led to the unnecessary increase in disease and premature death. The medical community has continued to produce more and more readily available medications and recommended procedures, many of which may not be helpful or may have harmful side effects. The individual diagnosed with serious illness is faced with spiraling healthcare costs which neither s/he or society can continue to meet. I am not here to debate the value of these treatments or to criticize our doctors, hospitals, or the drug and food companies that manufacture the medicines that we take and the foods that we eat. I leave that debate to others.

I am here to introduce a new path towards improving our lives, our health, and our future.

The Connexin Connection Molecular Fitness Lifestyle is not just another diet plan; nutritional fad; miracle cure; or one month plan to health, wealth and longevity. When one is 50 years old, 2,600 weeks of development have already taken place. So, why would anyone think that they could change their body and their habits overnight? *The Connexin Connection* is unique — it is based on a unified, scientific explanation for our existence and offers preventative and restorative solutions, a plan for optimum health, for the rest of your life.

The Connexin Connection offers a comprehensive health solution that requires focus, concentration, education, thought, and commitment, all aimed at helping you understand and appreciate your greatest asset, your health. Can you achieve it? If you have patience, perseverance, and time, life's greatest reward awaits you — good health for life.

There is nothing more important than health! The greatest minds throughout history have not agreed on many things, but they concur on health.

"To keep the body in good health is a duty, otherwise we shall not be able to keep our mind strong and clear."

Gautama the Budda, 563 BC

"A wise man ought to realize that health is his most valuable possession."

Hippocrates

"A healthy body is the guest-chamber of the soul; a sick, its prison."

Francis Bacon

"If you have health, you probably will be happy, and if you have health and happiness, you have all the wealth you need, even if it is not all you want."

Elbert Hubbard

"Without health, life is not life; it is only a state of languor and suffering."

Francois Rabelais

The Connexin Connection views disease not as some highly complex and mysterious process better left to the gods of medicine, but rather as a condition to be prevented through restoration and maintenance of our connexin function which is regulated by your diet and lifestyle. This revelation is your key to reclaiming your health, happiness and wealth.

It sounds simple, and it is. This critical pathway to disease has not previously been identified and embraced by the medical community. It is time for that to change.

"The next major advance in the health of the American people will be determined by what the individual is willing to do for himself."

John Knowles

"In order to change we must be sick and tired of being sick and tired."

Author Unknown

"The patient should be made to understand that he or she must take charge of their own life. Don't take your body to the doctor as if he were a repair shop."

Quentin Regenstein

The Connexin Connection discusses health, disease prevention and life extension, with a focus on connexin molecules that control the way the cells in our bodies work together. These molecules form the connections between our cells. They provide the means for communication between cells and the means by which our billions of cells can work effectively together as a unified whole. As such, *The Connexin Connection* program is a holistic approach to health, perhaps the most holistic approach ever advocated.

There is much scientific evidence that with excellent management of our life and health we will have open, effective, excellent communication between our cells. The result is superior health, protection from disease and the potential for the extension of life.

Empower Yourself!

Let's Summarize

- **Cellular Transformation** began some 600 million years ago with the appearance of the connexin molecule.
- **Species Transformation** resulted in the emergence and dominance of Homo Sapien over Neanderthal.
- Homo Sapien consumed a diet rich in connexin enhancing molecules.
- Our modern food products lack connexin enhancing molecules.
- Modern Man is now engaged in **Health Transformation** becoming more and more chronically ill, with no end in sight.

CONNEXINS 2

"Living bodies are even in the smallest of their parts machines ad infinitum."

Gottfried Wilhelm von Leibniz

"Man's mind, once stretched by a new idea never regains its original dimensions."

Oliver Wendell Holmes

The focus of this book is connexins — little known molecules that play an absolutely critical role in the prevention of disease and the maintenance of health. But in order to appreciate and understand the importance of these molecules and their pivotal role in health and disease, we need to understand their function in the human body.

Simply, life as we know it would not exist without connexins. The development of disease and death itself are directly related to connexin dysfunction. To put it another way — health and wellness are absolutely dependent upon our connexins functioning well if not optimally.

But what exactly are these molecules called connexins?

Connexins are proteins coded by our genes that first appeared in nature coincident with the transition of life from single cell organisms such as bacteria and amoebas to multicellular organisms such as ourselves.

So connexins appeared, and life as we know it evolved. *What is Life?*, written by Noble Laureate Erwin Schrödinger, in 1944, says, "Organic chemistry, indeed in investigating more and more complicated molecules, has come very much nearer to that aperiodic crystal which, in my opinion, is the material carrier of life." Could that molecule be the connexin molecule and, if so, what exactly does the connexin molecule do?

Imagine a sea full of single cell organisms all living relatively isolated lives and fending for themselves. **2-3 billion** years ago, these life forms had no means of communicating with each other and they had no ability to organize themselves in a meaningful sense. The ocean was teeming with life, but each organism was basically limited to what it could do as a single cell. The sum of the billions of single cell organisms equals not much more than an individual organism. So life continued on this way without much advancement for some 2 billion years.

For uncertain reasons, the connexin molecule made its appearance, and multicellular life arose on the Earth.

Connexins are found in all multicellular organisms and in almost all types of cells. A connexin is a protein composed of a sequence of amino acids as specified by a gene. Imagine a connexin molecule as a cylindrical rod available in 21 colors. Twenty-one types of connexin molecule subtypes are known and can be used to build a connexon (Figure 2.1), which is comprised of 6 connexin proteins. Two connexons, one in each cell membrane can join to form one connexin channel.

FIGURE 2.1 A Connexin

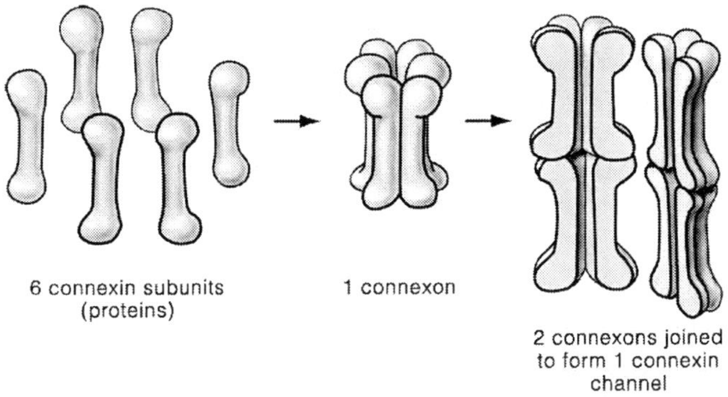

The connexon has a pore (or opening) at its center that can open and close depending on certain environmental conditions and the nutritional molecules it is exposed to (Figure 2.2). In figure 2.2 you are looking at the connexon molecule's opening or pore and as shown on the left the pore is open and on the right the pore is closed.

FIGURE 2.2 A Connexin molecule pore

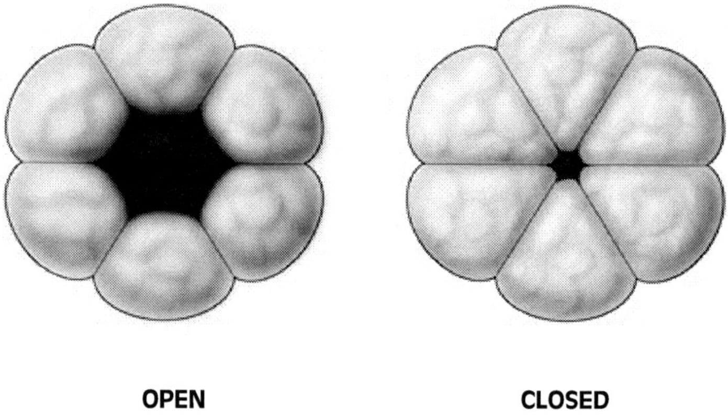

With 21 types of connexin proteins, each with slightly different properties, many types of connexons can be formed and used in different tissues, and for different purposes. Each connexon is inserted into the cell's membrane, as a window, porthole, or iris diaphragm is inserted into a home, ship, or camera, respectively. Now picture two cells, each with numerous connexons in their membranes (like docking ports on a space ship) coming together and their connexons lining up and docking together (see Figure 2.3).

FIGURE 2.3 Two cells with connexons docked forming connexin pores

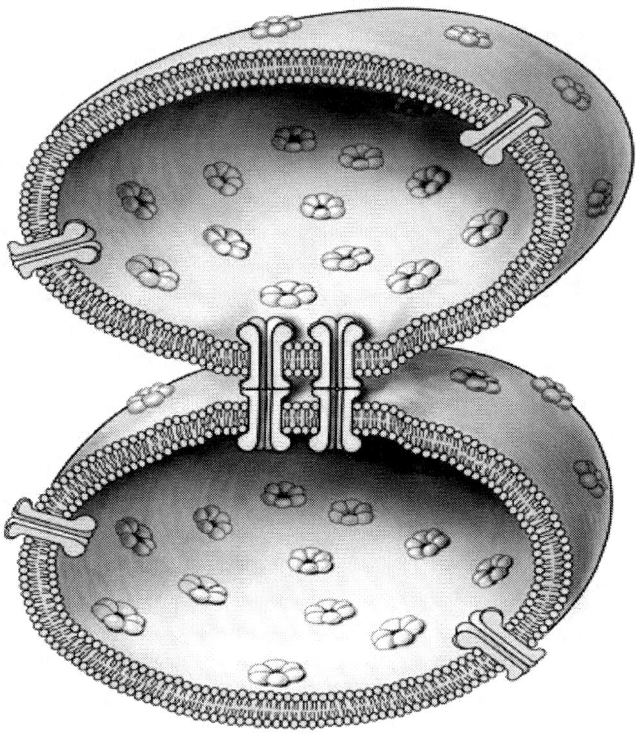

Once docked and under the right conditions, the central pore of each connexon can open, allowing the two cells to directly communicate. This assembly of two connexons docked together such that they allow a continuous pathway between cells is known as a gap junction. To reduce potentially confusing terminology, the term *connexin* will be used to refer not only to the connexin protein, but also to refer to connexons and the gap junction. Once this pore is formed, **flow** is established between cells. Now, the organism is greater than the sum of the parts and functions as a unified whole.

FIGURE 2.4 Molecular exchange through connexin pores

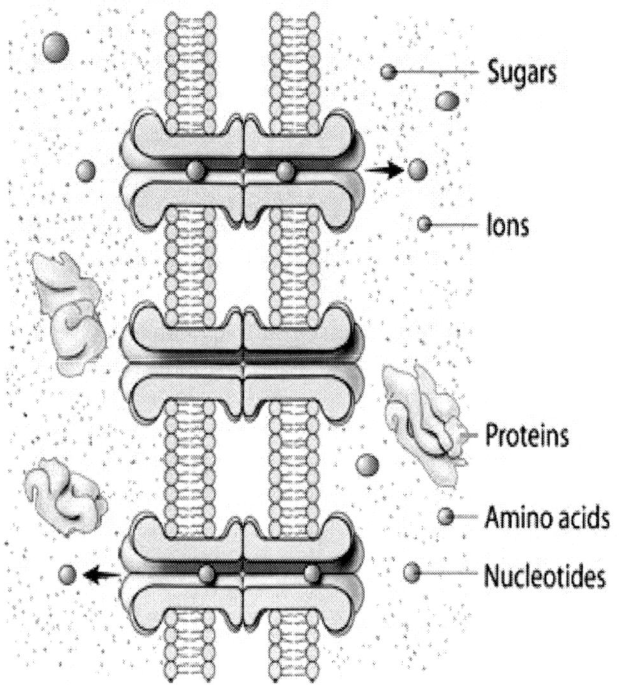

The connexin pore size is large enough for the exchange of the small molecules essential for cellular function, Figure 2.4. Yet, not all molecules, such as proteins, can pass through the hexagonal connexin opening which has a typical diameter of 6×10^{-8} inches (15 angstroms). Through open connexins, as seen in Figure 2.4, pass ions such as calcium, magnesium, potassium, sodium, zinc, and many others; cyclic nucleotides (used for cellular energy and regulation); amino acids (used to build proteins); sugars; nucleotides (the building blocks of genes); and a multitude of other molecules.

Not only are molecules being exchanged between cells; but when the connexins are open, cells are in electrical continuity with one another. This cellular continuity allows for cellular coordination and organization of similar cell types.

Connexins are involved in the regulation of all cells, integrating their function and thereby making multicellular life possible. Connexins are involved in controlling cell proliferation, type, and growth. Just imagine an organism that kept randomly growing and could not be stopped! The differentiation of cells (some into heart cells, skin cells, brain cells, muscle cells, etc) is critical to the formation of an integrated functional being. The activity of cells must also be quickly coordinated. Connexins play a role in both differentiation and coordination. Connexin based cellular communication is the fastest and most secure route for both electrical and molecular exchange. With connexin coupling our cells communicate on nature's wireless network. Even programmed cell death (apoptosis), which is essential for an organism's function, growth, and survival, is controlled in part by connexins. After an organism is fully formed, environmental changes occur and injury can take place. Once again, connexins play an invaluable role in the repair of tissues and in allowing cells to adapt to changes in their environment.

Connexin composition is not fixed, but rather changes in response to changing conditions.

The number of connexins per cell is variable; changing all of the time in response to our body's needs. Connexin number depends upon the cell type, the developmental stage of the organism, and the physical state of the organism at the time of measurement. Numbers can range from 10's to 1000's per cell. For example, in the cells of the uterine muscle during pregnancy, the cellular connexin number increases over 10-fold in preparation for the uterus to progress from a non-contracting organ to a fully contracting one during labor. During in-utero development of a fetus, connexin numbers change in different cell types as organ development proceeds toward a fully complete being. The number of connexin molecules on a cell can vary regionally and be very densely packed into an orderly array.

Wound healing and tissue repair, vital to all of us, are processes in which connexin number changes in response to an injury, playing an important role in the restoration of damaged tissues.

To visualize connexin molecules creating multicellular life by joining individual cells together, look at Figure 2.5. The image of a woman is formed from a collection of individual spheres placed in proximity in Salvador Dali's 1952 painting entitled

Galatea of the Spheres. However, it is an image from unconnected spheres that have no life. A human can be created if these spheres are actually billions of cells in proximity connected by connexins.

FIGURE 2.5 Dali's Galatea of the Spheres

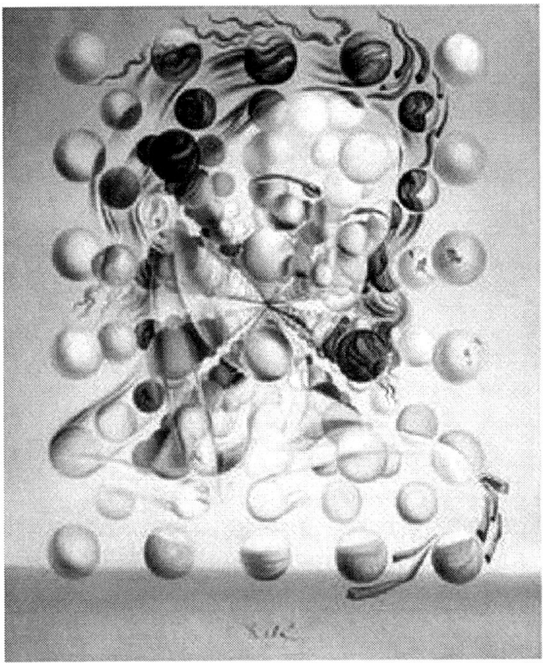

Now, imagine ourselves as a Sand Sculpture where the spheres are tiny sand grains (Figure 2.6, Courtesy of www.TeamSandtastic.com) which are much closer in size to our cells, but still many times larger. The fully formed Sand

Sculpture may look almost as life-like as a human, but still remains an inanimate object.

FIGURE 2.6 Sand sculpture

Our cells, unlike sand grains, are alive and when fully connected can function together as a unified organism — they are connected physically, molecularly and electrically by connexins. Fully connected living cells together become greater than the sum of the individual cells — they become life as we know it. Health is a state of connectivity-wholeness. Disease is a state of disconnection and disorganization — falling apart.

In the following chapters we will learn how the 8 major diseases of modern man are all related to connexin dysfunction, and how we can embrace a connexin lifestyle to both prevent disease and restore excellence of connexin function.

Let's Summarize

- Connexins are proteins coded by our genes.
- The cells of multicellular organisms communicate via connexins.
- Connexins are involved in the regulation of cellular processes and the integration of cells.
- Connexin number is variable and changes in response to our body's needs.
- Connexins allow the trillions of cells in our bodies to function as a unified whole by connecting them physically, molecularly and electrically.

CONNEXINS AND DISEASE 3

"Unless we change direction, we are likely to end up where we are headed."

Chinese Proverb

"The major advance in the health of the American people will be determined by what the individual is willing to do for himself."

John Knowles

CONNEXINS AND DISEASE

Before we learn about the relationship of disease to connexin function let's discuss what constitutes disease. Today in America and throughout most of the developed world disease is an everyday word. The cocktail party conversation often revolves around how many diseases you and/or your friends have. Illness has become the status quo, but what is disease?

To begin understanding disease let us differentiate between **extrinsic disease,** like a bacterial infection, versus **intrinsic disease**, like the eight diseases of modern man: cancer, heart disease, diabetes, arthritis, osteoporosis, Alzheimer's, hypothyroidism and ocular disease.

Medicine "took off" and "took hold" because of its initial success treating extrinsic disease like bacterial infections with penicillin in the 1940's. Today, man continues to have infinite hopes for miraculous cures of intrinsic disease because of his success with killing bacteria. The average man and even physicians continue to misunderstand that treating a bacterial infection by killing foreign bacteria is not a good solution for treating intrinsic disease. Treating intrinsic disease involves our own cells, where prevention or restoration of cellular function is the preferred solution.

We need to stop and think about why the treatment of extrinsic disease is vastly different than treating intrinsic disease. Treating infectious disease is basically no different than using a pesticide on your lawn or garden to kill organisms that are attacking your grass, fruits or vegetables. Antibiotics are like pesticides; they are developed to poison and kill offensive foreign organisms (bacteria) without harming human or plant cells. Antibiotics do not improve your body's immune system or benefit human cells directly. Antibiotics kill bacteria so that the bacteria do not kill us. This process of killing nonhuman life forms is based upon the fact that they are physiologically different enough from us that we can chemically target a unique aspect of their physiology that results in their death, without hurting ourselves. To treat or prevent an intrinsic disease, like one of the eight diseases of modern man, requires maintaining or restoring cells to a state of molecular fitness thereby optimizing cellular function and health.

Generally, external diseases can be prevented. Disease causing bacteria only gain access to our bodies because we have damaged our body by trauma such as a cut or created a situation where the coverings of our body, both inside (endothelial cells — for example, the cells lining blood vessels) and out (epithelial cells — for example, skin), are defective because of poor cellular connectivity. When this happens harmful bacteria have easy access to our bodies. An analogy of bacterial invasion is the attack of a gated castle. If all of the bricks of the castle are in place and reinforced with good mortar and there are no open entries into the castle, then the foreign invaders cannot enter and kill the inhabitants. Similarly, if we consume a diet rich in connexin enhancing molecules all of our cells will be well connected and communicating and bacteria will have a difficult time entering our body. Our poor nutritional intake contributes to a lack of connexin enhancing molecules and thereby contributes to the development of recurrent and poorly healing skin infections, upper respiratory infections (colds) and bladder infections. We should take a preventative and restorative approach to infections by enhancing our connexin function and correcting any nutritional deficiencies we may have that would lead to thin and poorly connected skin. Conventionally, we are likely to be placed on single dose or chronic antibiotic therapy.

Unfortunately, having had such good success in treating extrinsic diseases such as bacterial infections with antibiotics, modern medicine has thought that the answer to treating all disease was to kill it. Medicine has adopted the paradigm of "war against disease," where disease has been treated by either cutting it out or by killing our own human cells rather then repairing those cells. Our treatment of cancer is based upon killing cancer cells with chemotherapy, surgery or radiation. All three modes of treatment also destroy and impair the function of healthy human cells in the process. The hope in treating cancer is that the treatments themselves will kill the cancer cells before they kill you, but unfortunately this is not always the case and sometimes the treatment kills the patient. The medical profession does not seem to accept that the reason cancer develops and metastasizes is because our cells have been placed in such a poor environment, a connexin adverse environment, and rather than work together our healthy cells choose to transform into cancer cells and make a go of it on their own. If we restore our body back to the conditions it was designed to function best under, and restore our cells to be made out of the molecular constituents optimal for them, they would no longer be cancerous, but rather they would be healthy, unified, functional parts of our human body.

Our concept of killing disease is so primitive that it defies all reason and logic. For example, if you want to eradicate the world of terrorists and prevent the formation of new terrorist groups in the future, does attacking and killing a single terrorist group accomplish this? Doubtful. Killing a single terrorist group in the short run may solve the immediate problem, but in the long run one needs to address the reason why terrorist groups exist at all. To prevent future terrorist attacks one needs to create a worldwide environment that does not breed, encourage or foster the genesis of future terrorists. This situation is identical to disease states in man. If one wants to prevent disease or prevent it from recurring after it has been treated, then one needs to optimize the environment of their cells so disease will not occur again nor will new diseases be encouraged to develop.

Certainly, the eight intrinsic diseases of modern man were never as common as they are today. Some believe reasons for increased disease to be that man is living longer today and that modern medicine has sophisticated diagnostic tests that can detect what was previously undetectable. These points, on the surface, do seem to offer reasonable explanations for the increased morbidity and mortality resulting from the eight major diseases. But maybe these explanations are nothing more than a rationalization, an excuse, for the "modern" lifestyle we are living today.

Yes, it is true that, on average, man lives longer today. However, over 90 percent of this average gain in life is attributable to improvements in infant mortality; children under 1 year of age are living longer. You as an adult are not enjoying more years of good health. The explanation for this average increase in life expectancy is not the eradication of the eight major diseases of modern man. Rather, improvements in sanitation, reduction of infectious disease, and decreased child mortality are the more significant factors. What is truly different today is illness. Never has man lived as ill as he does today. The average American will spend some 13 to 14 years before death in a state of illness, taking typically 10 or more prescription medications. Our quality of life has been diminished.

As long as 2000 years ago, some people were recorded as being functional, creative, writing books, and doing so well into their 90's and 100's. This is documented in many of the writings of the pre-Socratic philosophers and scientists; even the father of medicine, Hippocrates, lived well into his 100's. I believe their diet and lifestyle were better than most have today.

Why is disease so common, why are so many people suffering from more than one disease and why are people developing more of these diseases as they get older? I believe there is a simple answer to all of these questions — **diet and lifestyle choices leading to connexin dysfunction.**

That is why first and foremost, it is important to understand that each of the eight major diseases of modern man are related to connexin dysfunction.

Table 3.1 summarizes the results of hundreds of scientific studies looking at connexin function and dysfunction as related to the eight diseases of modern man. In every case, the incidence of disease has been shown to be accompanied by impairment of connexin function.

TABLE 3.1 Connexin function and disease

1 Cancer	2 Heart	3 Diabetes	4 Arthritis	5 Osteoporosis	6 Alzheimer's	7 Hypothyroid	8 Ocular
↓Connexin Function	↓Connexin Function	↓Connexin Function	↓Connexin Function	↓Connexin Function	↓Connexin Function	↓Connexin Function	↓Connexin Function

The probability of this type of association occurring by chance in eight independent diseases (as they are generally considered to be by the medical profession) is 1 in 256. These statistics imply a very high probability that all eight of these diseases are linked to the connexin molecule. Most medical and scientific studies consider a result to be statistically significant if the probability of occurring by chance is 1 in 20 or less.

Based upon this statistical analysis, and all of the research that has gone into producing this scientific data, it is time for re-evaluation of our current medical paradigm. Rather than consider disease to be organ or tissue specific and that the prevention and treatment of disease should be organ or tissue specific, we need to unify these 8 diseases and their prevention and treatments.

The Connexin Connection is a composite of several thousand years of scientific and medical research. Connexin research is relatively new; yet, encompasses four or five decades of research. This book and the premise of disease integration is derived from a thorough collection of medicine, physiology, biochemistry, molecular biology, biology, nutrition, physical medicine, evolution, monitoring, and philosophy.

The way to think about this book's composition is to visualize a puzzle or a digital picture made out of hundreds of thousands of pieces or pixels. If you were to look at the pieces or pixels in isolation, there would be no picture. As the number of pieces grows or the pixel density increases an image begins to form. Then, as the pieces or pixels become numerous enough our brain forms a coherent image from the seemingly meaningless bits of information. Finally, it all becomes clear! I hope that the information and concepts presented in *The Connexin Connection* coalesce into a clear and meaningful picture and that the whole becomes greater than the sum of the parts.

Connexin research seems new, but not so — connexin research dates back to the 1960's and is well established in over 5,000 peer reviewed scientific papers. This is substantial in terms of both fields of study involved and technologies used to investigate connexins. This research has been confirmed and validated by numerous major scientific laboratories at the best institutions throughout the world. The complete gene and amino acid sequence is known for connexin molecules. Their composition and structural arrangements in cell membranes are known along with the consequences of their absence, as demonstrated by removal or selective "knock out" of connexin molecules in animal models.

The defects in connexin molecules or their alteration or absence in disease have been carried out in tissues from normal or diseased animals or humans. Studying connexins is not as simple as measuring something in the blood. Connexins are part of your cells and their function can only be evaluated in cells or tissues. Connexins are tissue and condition dependent. Meaning, they change throughout your life; beginning when you are first conceived, as you age, become ill, and one day die. **Connexins are a reflection of your life, and your life is a reflection of your connexin molecules.**

Connexin dysfunction or diminished connexin function can occur in several ways:

1) Reduced numbers of connexin molecules on cells (down regulation of genes or gene products — decreased gene expression).

2) Connexin molecule closure.
3) Alteration of connexin molecule gating properties.

These dysfunctions can result from diet, lifestyle or environment. Alternatively, molecules obtained from our diet and lifestyle choices can enhance connection function. These dietary and lifestyle factors will be discussed in detail in Chapters 4 through 10.

Once this association between diet, lifestyle and disease is clear, it becomes apparent that the current medical paradigm of classifying and treating diseases by relating them to separate organ systems is in need of reevaluation. I believe all of these 8 diseases are one disease that is related to diminished connexin function, although not exclusively.

I believe the time has come for implementation of a new medical paradigm that will focus on disease prevention and return man to wellness and health. I believe that new paradigm to be the prevention of connexin dysfunction and the restoration of connexin function.

In order to best understand why connexin dysfunction leads to disease we need to revisit the role of connexins in multicellular organisms — life as we know it. As discussed in Chapter 2, connexins are involved in the regulation of all cells and integrating their function by controlling cell proliferation, cell growth, cell differentiation (some into heart cells, skin cells, brain cells, muscle cells, etc), cell communication, and cell coordination.

The number of connexins per cell is variable; changing all of the time in response to our body's needs as well as diet, lifestyle and environment. If one of these connexin functions or the connexin molecule itself is made dysfunctional, a state of cellular, organ, and tissue dysfunction occurs — this we call disease.

To follow, I will describe in general terms the magnitude of the eight major diseases along with their relationship to the connexin molecule. Table 3.2 shows results from a number of studies that relate particular diseases to the types of connexin molecules that display dysfunction in them. While, as previously noted, there are 21 types of connexin molecules, the scientific numbering system does not follow a 1-21 numbering system (connexin is abbreviated "Cx").

TABLE 3.2 Disease and connexin specific dysfunction

Heart Disease / Cardiovascular Disease	Arrhythmia – Cx40, 43 Hypertension – Cx40 Atherosclerosis – Cx37, 43 Congenital malformation – Cx43
Cancer	Liver – Cx43, 32 Prostate – Cx43, 32 Brain (gliomas) – Cx43 Bladder – Cx26 Breast – Cx26, 43 Ovarian – Cx43 Kidney (renal cell) – Cx32 Chemical or Radiation Induced – Cx32, 43
Diabetes	Cx36, 43
Arthritis	Cx43
Osteoporosis	Cx43
Alzheimer's	Cx43
Hypothyroidism	Cx43, 32, 26
Ocular Disease	Cataracts – Cx43, 50 Congenital Cataracts – Cx46, 50
Peripheral Neuropathy	Cx32
Hearing loss	Cx26, 30, 31

HEART DISEASE / CARDIOVASCULAR DISEASE

Cardiovascular disease remains the number one killer of Americans. As of 2002, 34.2 percent of Americans were estimated to have cardiovascular disease, with 927,448 of them dying from this disease. There is death every 34 seconds from this disease, 2,600 Americans dying each day. In 2002, an estimated 6.8 million inpatient cardiovascular operations and procedures were performed in the United States alone. For 2005, the annual estimated cost of cardiovascular disease is projected to be $393.5 billion.

Connexins (Cx) are associated with many forms of heart disease; ranging from developmental heart defects in newborn children to adults, hypertension (high blood pressure), arrhythmias (irregular heart beats), sudden cardiac death, and atherosclerosis (hardening of the arteries).

There are predominantly three types of connexin molecules known to exist in the heart and blood vessels. These connexins play essential roles in normal hearts, and alteration of these molecules contributes to abnormalities that occur in human cardiovascular disease. Defects in specific connexins lead to distinct congenital heart defects. Arrhythmia is associated most consistently with a reduction in connexin 43 while hypertension may involve alterations in all connexin types expressed in arteries. Specifically, removal of Cx40 has been shown to be associated with hypertension and altered blood vessel regulation. Atherosclerosis has been shown to involve alterations of connexins 43 and 37.

Connexins play a critical role in heart disease.

CANCER

Cancer is projected to kill 570,280 Americans by the year 2005 — more than 1,500 each day. The lifetime risk of cancer for men in the U.S. is slightly less than 1 in 2; for women the risk is a little more than 1 in 3. It is estimated that 1,372,910 new cancer cases will be diagnosed in 2005. Spending on cancer cost our nation $189.8 billion in 2004 alone.

A role connexins play in cancer was proposed as long as 40 years ago. Since that time literally hundreds of scientific papers have validated this claim. Cancer cells are generally deficient in functional connexins. Many tumor causing agents are known to inhibit connexin function; while known cancer preventing agents, such as carotinoids, enhance connexin function. Both the development of and metastasis or spread of tumors involves connexin dysfunction. The list of tumor types associated with impaired connexin function grows each day and currently includes: prostate, brain (gliomas, astrocytic tumors), lung, breast, liver, bladder, ovarian, uterine (endometrial), and thyroid cancer.

Many of the known cancer preventative foods contain molecules that boost connexin number and function: carotinoids (carotene, lycopene), isoflavones (genistein), polyphenols (resveratrol, curcumin, and others). These will all be discussed specifically in later chapters.

Clearly, connexins play a huge role in cancer.

DIABETES

In ancient Greece, diabetes was known to exist, but it was a rare disease. Today, that has dramatically changed as diabetes now affects some 18.2 million Americans. Diabetes cost our nation $132 billion yearly as of 2002, a cost of $13,243 per affected person.

The insulin producing cells in the pancreas are connected by connexin 36. Changes in Cx36 alter the insulin content and glucose-induced release of insulin in these pancreatic cells. Based upon scientific studies, it is thought that the insulin secreting cell (beta cell) anomalies in both type I and type II diabetes are related to connexin dysfunction. In addition, many of the side effects of diabetes such as: diabetic cardiomyopathy, arrhythmias, diabetic retinopathy, diabetic peripheral neuropathy, diabetic cataracts, and diabetic microangiopathy may all be related to connexin dysfunction.

Diabetes and connexin dysfunction go hand-in-hand.

ARTHRITIS

Nearly 70 million Americans, or about one in three adults, were affected by arthritis in 2001, at a cost of $82 billion annually. The prevalence of arthritis increases with age, with approximately 60 percent of people over 65 years of age suffering from this disease. The number of people over 65 with arthritis is expected to double by 2030.

Arthritis can be initiated by inflammation of the joints which can then progress to connexin dysfunction. Key tissues that are involved in arthritis line the joints, synovial cells and cartilage cells (chondrocytes). When healthy, these cells are connected by connexin 43 and a smooth intact lining is formed. Inflammatory molecules known to produce arthritis have been shown to close connexin 43 junctions leading to disruption of these tissues.

Yes, connexins and arthritis are linked too.

OSTEOPOROSIS

Osteoporosis and low bone mass are currently estimated to affect 44 million Americans, with that figure rising to 61 million by 2020. Some 55 percent of people over the age of 50 are diagnosed with osteoporosis. Osteoporosis is responsible for more than 1.5 million fractures annually at a cost of $17 billion dollars in 2001.

Connexin 43 is the major connexin in bone cells, osteocytes and osteoblasts. Cellular communication via these connexins is essential for the health of bone cells, and the transfer of mechanical signals among them which makes for strong healthy bone. Parathyroid hormone is known to be critical for bone formation and increases both Cx43 levels and their function; whereas a known precipitator of osteoporosis, metabolic acidosis, closes connexins.

Even bone health depends on connexin function.

ALZHEIMER'S DISEASE

An estimated 4.5 million Americans have Alzheimer's disease, this number having doubled since 1980. By 2050, the number is expected to reach 16 million. One in 10 Americans has a family member with Alzheimer's and one in three knows someone with the disease. The average Alzheimer's patient will live eight years with the disease and lives on average 20 years from the first onset of symptoms. The average lifetime cost for an individuals Alzheimer's care is $174,000, with the national annual cost being over $100 billion.

The cause of Alzheimer's disease remains unknown. Nonetheless, alterations in connexin function have been implicated in this disease too. Recently, a regulator of connexin 43, a molecule called drebrin, has been shown to be reduced in Alzheimer's brains. The result of this reduction in drebrin is reduced cellular communication due to a coincident reduction of connexin 43.

Are you seeing the pattern yet?

HYPOTHYROIDISM (OBESITY)

Hypothyroidism was estimated to affect 13 million Americans in 2000. This condition is associated with weight gain, inability to lose weight, fatigue, depression, decreased libido, and high cholesterol.

A low-iodine diet results in hypothyroidism. One of the effects of low iodine is a reduction in thyroid cell connectivity by connexins. In autoimmune (immune system related) thyroid disease, which results in hypothyroidism, the diseased tissue has reduced amounts of connexin 43, connexin 26, and connexin 32. The hormone, thyroid stimulating hormone (TSH), produced in the hypothalamic region of the brain, regulates thyroid hormone production by increasing the connexin function of thyroid cells.

Connexins also play a role in this major disease.

OCULAR DISEASE — CATARACTS / MACULAR DEGENERATION

Cataracts affect 17 percent of Americans over age 40. Projections indicate 30 million people will be affected by 2020. An estimated 1.5 percent of Americans over age 40 have macular degeneration and researchers predict this number will increase to three million by 2020.

The cells of the lens in the eye are connected predominantly by connexins 46 and 50. Cataracts are caused by disruption of the lens cells, resulting in the loss of lens transparency. Congenital cataracts as well as the development of cataracts in the aging population are associated with a reduction in connexin function.

A number of other eye disorders involve reductions of connexin function in the retina (diabetic retinopathy, macular degeneration), lacrimal gland (dry eye from decreased secretions) and cornea.

Eye health is a matter of connexin function.

SUMMARY

As you have learned, connexin dysfunction is a key feature of all of the eight major diseases of modern man (see Table 3.1). There is a very high probability and real likelihood that all of these eight diseases are linked to the connexin molecule and its function.

I believe these major diseases are a singular disorder related to connexin dysfunction, which can be treated and/or prevented by interventions directed at restoration of connexin function.

The following chapters will concentrate on how you can optimize your connexin function through diet, exercise, and stress reduction — specifically by adopting **The Connexin Molecular Fitness Lifestyle.**

Let's Summarize

- Decreased connexin function is found to be present in the eight major diseases of modern man.
- Some congenital diseases (birth defects) are also associated with abnormalities or the absence of specific connexin subtypes.
- It is statistically remarkable that all eight major diseases are associated with a single common disorder, connexin dysfunction.
- Could all of our major chronic diseases actually be one disease resulting from changes in our ancestral diet and lifestyle?

4 Connexins and the Food We Eat — I

"When diet is wrong, medicine is of no use. When diet is correct, medicine is of no need."

Ancient Ayurvedic Proverb

"We are always getting ready to live, but never living."

Ralph Waldo Emerson

CONNEXIN CLOSERS / FUNCTION REDUCERS → DISEASE / DEATH

What we eat is critical to the function of our connexin molecules. If we eat the proper foods, our connexin molecules remain open and connected and operate so that we will be healthy. If we eat the wrong foods, our connexins become dysfunctional and closed, and we may become ill.

As discussed in the introduction, a connexin enhancing diet may have been responsible for modern man's (homo sapien) ascent to the top of the animal kingdom. However, this diet has changed radically since our emergence around 100,000 years ago.

Over the last 60 years, the creation of processed foods has accelerated dietary change as never before! Simply put, the human genome can not possibly adapt quickly enough to such a rapid and radical alteration of the food supply. Our biochemistry and physiology are left confused and disrupted by this new diet. Humans are no different than all other species in that they are genetically adapted to their ancestors' environment and as such are vulnerable to rapid environmental change. As with other rapid and radical environmental changes, such as the Ice Age, the full catastrophic consequences of dietary change have taken many generations to fully manifest. And now, we have become the sickest society to have ever lived on Earth.

Our modern processed-food diet has resulted in a major reduction of molecules in our food that can open connexins, in favor of molecules that close or reduce connexin function. Selenium is an example of a connexin enhancer once abundant in the soil, now depleted due to agriculture. Selenium depletion has been associated with cancer and for this reason is replenished in the soil of Scandinavian countries. Today, only foods such as Brazil nuts, grown in the mineral rich soils of the Amazon contain significant concentrations of selenium.

In this chapter, foods in our diet that reduce connexin function will be presented with respect to 5 fundamental nutritional characteristics of the ancestral human diet as compared to our current modern diet: 1) acid-base (pH) balance, 2) sodium/potassium and chloride/bicarbonate ratio, 3) macronutrient composition, 4) glycemic load, and 5) fatty acid composition (see Figure 4.1).

FIGURE 4.1 Connexin Closers / Function Reducers

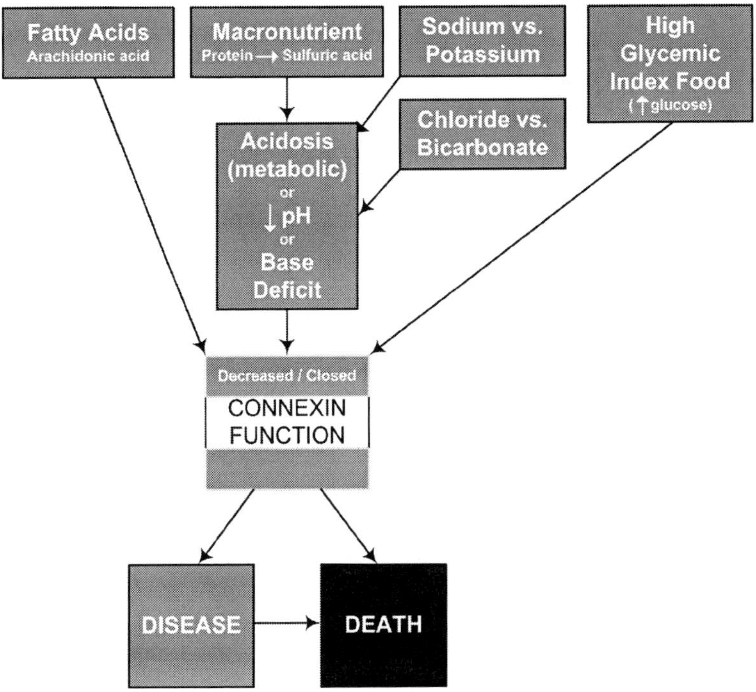

GENERAL INTRODUCTION TO ACID / BASE CHEMISTRY IN OUR BODY

The human body and most life forms strive not to be acidic; they prefer to be the opposite, or slightly alkaline. Clinically, a blood pH of 7.4 is considered normal. If your blood pH is less than 7.4 you are said to be acidic, have an acidosis, or to have a base-deficit; whereas, if your blood pH is greater than 7.4 you are termed alkaline, said to be alkalotic or to have a base-excess.

Briefly, an *acid* is a substance that when dissolved in water produces hydrogen ions (H^+). A *base* is a substance that produces hydroxyl ions (OH^-) in water. When combined in equal amounts they neutralize each other to produce water: $H^+ + OH^- = H_2O$. The amount of acid or base in a solution is generally reported on the pH scale, where $pH = -\log [H^+]$. When hydrogen ions are in high concentration, the solution is said to be acidic, and the pH low. When the pH of a solution equals 7 it is said to be neutral. A pH less than 7 identifies a solution as being acidic, and a pH greater than 7 identifies the solution as basic or alkaline. I mention the pH scale because it will be referred to throughout this book. Commercially available pH strips can be used by you to easily measure and monitor your degree of dietary acid production. **Please note that pH is measured on a "log scale" which means that each 1-point change in pH is actually a 10-fold difference in acidity or alkalinity.** For example, the reduction in acidity between a pH of 5 and 6 is not 1, but 10 times! In a similar fashion, the reduction in acidity between a pH of 5 and 7 is not 2, but actually 100 times! (10 x 10). Likewise, the difference in acidity between a pH of 5 and 8 is not 3, but 1000 times! (10 x 10 x 10).

Clinically, the gold standard for measuring a person's overall acid / base balance is blood pH. Blood pH is an overall composite of what is going on in the whole body and is regulated primarily by your kidneys, lungs, intestines, and liver. To know exactly what is going on at the cellular level would be ideal, but it is not clinically practical to measure the pH within the cell. Practically, the most useful pH measure for following one's diet and consumption of acid producing or alkalinizing foods is to measure urine pH.

Different parts of our body (organs, blood, urine, other secretions, etc.) have pH values that differ in response to our diet, yet they are all interrelated as shown in Figure 4.2. It is important to understand that urine pH, blood pH and cellular pH all change together in the same directions over time. Blood pH is least sensitive to change on an hourly or daily basis because it is well buffered or regulated. Because of this tight regulation, small pH changes occurring in the blood, even 0.1 pH unit, can signify a major abnormality or illness.

Urine pH can change quickly and by large amounts (2-3 pH units) in response to a person's diet. This is due to the net acid or alkali generating properties of one's food. Thereby, urine pH provides you with rapid feedback regarding the nature of your diet. A rapid change in urine pH does not signify that a rapid change in blood pH will follow. However, if the urine pH is constantly measured as acidic over long periods of time (months) it is indicative that the body is chronically producing large amounts of acid (most likely from acid generating foods). In this situation, the blood and cells themselves eventually become slightly acidic (low-grade metabolic acidosis), and this cellular acidosis can result in connexin closure and impairment, leading to disease. On the other hand, if your urine pH is constantly measured as alkaline over time this is an indicator that your diet is overall net alkalinizing, and your body is not likely to be acidotic.

Figure 4.2 begins with acid producing food or dietary constituents being broken down by your intestines and then transported through the blood to your liver and other cells where they are metabolized. Metabolism of protein and organic acids by the liver results in acid generation (H^+) which enters the blood and eventually reaches all cells. Other constituents in our food are absorbed and metabolized to yield acid (H^+) or base (OH^-).

In this figure two tissue types, bone and muscle are shown in continuity with blood. When the blood is acidic (H^+) and equilibrates with muscle this leads to muscle breakdown.

FIGURE 4.2 Effect of acid forming foods

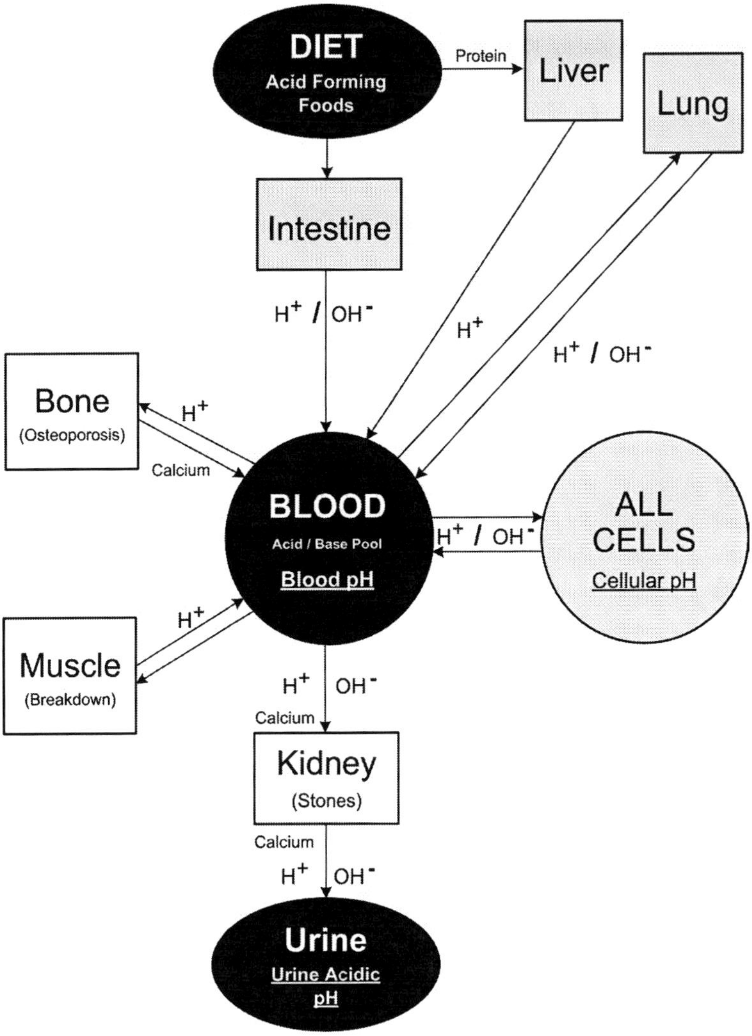

Acidosis of the blood results in calcium removal from the bone that is used to buffer the blood or in other words neutralize the acid. This loss of bone calcium results in osteoporosis. In turn, the calcium in the blood is filtered by the kidney and can lead to kidney stone formation.

Our urine pH reflects the net renal acid load generated in our body, which comes primarily from eating acid generating foods. Therefore urine pH is highly reflective of our diet, acidifying or alkalinizing.

CONNEXIN CLOSURE BY AN ACIDIC PH

As previously discussed, the eight major diseases are linked to the connexin molecule and its dysfunction. Medical and scientific studies show that the net acid producing diet of modern man along with declining kidney function (this reduces the ability to remove acids from our body) as we age, results in the formation of a metabolic state known as chronic low-grade metabolic acidosis. This acidosis — too much acid in our body can diminish connexin function and lead to disease.

Connexin function is regulated by many molecules and conditions; pH — the degree of acidity or alkalinity — is one of the more important ones. This pH-regulation of connexins applies to all cell types studied thus far including: bone cells, eye cells (cornea, lens, and retina), brain cells (neurons), heart cells, skin cells, and even to connexin molecules expressed in the laboratory. In this section, closure of connexins by an acidic pH will be presented. In the next chapter, opening of connexins in response to an alkaline pH will be discussed.

In addition to opening and closing the connexin molecule, pH also regulates connexin gene expression or the number of connexins. An acidic pH reduces connexin mRNA (the message from which connexins are made). This results in decreased production of connexin molecules, which in turn results in a net decrease in overall cellular connexin function.

In an acidic environment connexins are closed and/or their function is reduced. An example of how profoundly changes in pH can affect connexins is clearly demonstrated in bone cells. At a pH of 6.9 only 40.7% of bone cells are connected; however, within 1 hour of changing the pH to 7.6 an overwhelming 92.8% become connected. This regulation of the connexin molecule by pH, as illustrated by bone cells, demonstrates the profound effect of acidosis on cellular connectivity. No surprise then that acidosis results in osteoporosis.

Now just imagine what happens when your entire body's pH is reduced to 6.9. At this pH level your body is 60% disconnected! This state of being is not healthy. In fact, if allowed to continue it leads to death.

As a physician sitting at a dying patients' bedside on many a night, I watched and pondered death as it gradually unfolded.

The waning hours of life reflect connexin closure perfectly. A patient who was once alert and talking gradually becomes confused as they become acidic and connexins in the brain close. Soon, the patient is no longer responsive. As the cells lining blood vessels and smooth muscle cells in their walls disconnect, the arteries dilate and blood pressure drops. Regular heartbeats become more and more irregular as the heart's electrical conduction system disconnects. Breathing changes into strange patterns as the cells controlling respiration disconnect from one another. Bowel sounds stop as cells in the intestines disconnect. Finally, upon reaching a blood pH of around 6.9, the heart beats irregularly. The monitor screen soon shows a flat line. The patient is dead!

I sought to relate the physiology and biochemistry I had been taught in medical school to this process called death. I was watching the dismantling of life, yet I was totally helpless in restoring function, and there was no one to explain this mysterious process to me. Now, many years later, having developed an understanding of life in terms of pH and connexins (cellular connectivity) this process begins to make sense.

In an intensive care unit, modern medicine institutes the ritual of blood pH measurement. Doctors watch patients' blood pH dropping, becoming more acidic, and frantically try to restore the pH number in hopes of sustaining life. When injury or disease is too advanced and one is unable to correct the underlying problem, you hope that the body will repair itself or that a miracle will occur. As you monitor the patient's blood pH, it becomes more and more acidic, and the patient gradually disconnects to a point where he or she is pronounced dead. Acidosis or low pH is a great predictor of patient death. In fact, acidosis has a positive predictive value for death of 96%!

Acidosis during life can occur as a result of diet. The dietary changes which started 10,000 years ago and which have greatly accelerated over the last 60 years have caused successive generations to become more and more acidic. Today, most Americans reside in a state of low-grade metabolic acidosis, have impaired connexin function, and are at risk of contracting one or more of the eight diseases of modern man.

DIETARY INDUCED CONNEXIN DYSFUNCTION

As seen in Figure 4.1, the key dietary changes responsible for connexin dysfunction in our modern diet are:

1) Addition of chloride to our food, mainly in the form of sodium chloride (salt, sea salt), but more recently in the form of potassium chloride (salt substitutes).

2) Dietary reduction of bicarbonate-generating molecules in our diet that naturally neutralize acid because of reduced consumption of fruits and vegetables.

3) High Protein diets that produce sulfuric acid from metabolism of sulfur containing amino acids in protein, especially prevalent in animal protein.

4) Increased arachidonic acid intake from food. Arachidonic acid is a fatty acid derived primarily from a diet rich in animal fat.

5) Processed (refined) grain and sugar products that have higher glycemic loads than natural whole grains or sugars.

These connexin closing nutritional elements result in net acid generation in our bodies. Recognition of dietary acid production and calculation of the acid and base forming properties of foods is not new. In 1907, Sherman and Sinclair first published these calculations. Yet, in 2005 most physicians and nutritionists fail to recognize the importance of our body's acid / base balance in relation to diet and its impact upon disease and life itself. Remer and Manz in 1995 published a detailed list of foods and their acid or base generating potential. Commonly consumed food groups resulting in net acid production, called PRAL (potential renal acid load or the amount of net acid generating potential of a food) are shown in Table 4.1.

TABLE 4.1 Major food groups that are net acid generating

Average PRAL per 3.5 ounces of a food

Food Group	PRAL
Milk and non-cheese dairy	0.95
Soft Cheese	4.51
Bread, Grains	5.73
Eggs	8.21
Fowl and Fish	8.45
Meat	8.38
Meat Processed	11.42
Cheese Hard	19.22

DIETARY ACID PRODUCTION

SODIUM / POTASSIUM AND CHLORIDE / BICARBONATE

Yes, it is true — "salt" is bad. Henry David Thoreau wrote, "Finally, as for salt, that grossest of groceries, to obtain this might be a fit occasion for a visit to the seashore, or, if I did without it altogether, I should probably drink the less water. I do not learn that the Indians ever troubled themselves to go after it." We have greatly increased our "salt" or sodium chloride (NaCl) intake by the addition of various sodium chloride sources into our food (salt, sea salt, etc.). This increased sodium intake has resulted in the reversal of the sodium/potassium ratio our body was accustomed to. Scientific evidence indicates that our ancestor, hunter-gatherer man, ingested a diet containing three potassium molecules for each sodium molecule in his diet. Today, the average American diet is so radically changed that we consume ten times as many sodium molecules as potassium molecules! This increase in sodium is not the real problem — chloride is. Along with each sodium molecule we consume, we also gain an additional chloride molecule. Chloride leads to acidification of our body. And if all this **sodium chloride** is not bad enough; more recently, **potassium chloride** and other salt substitutes have been introduced into our diet, furthering our chloride intake.

Unlike today's salt addicted man, ancient man had the sense to abstain from drinking sea water or brackish water and did not actively mine inland dry sea beds for sea salt. Artificial addition of sodium chloride into our diet began about 2,000 years ago in China as a major revenue source for the Chinese Emperors. They soon discovered this substance was highly addictive to the populous. Upon visiting China, our European forefathers quickly saw the financial benefits of introducing salt into the European diet, and this practice continues throughout the world today.

In addition to reversing our dietary sodium / potassium ratio, the modern diet has also reversed our dietary chloride / bicarbonate ratio. Nature created a solution for combating dietary acid generation in the form of all fruits and vegetables, rich in bicarbonate-generating molecules, which neutralize acid produced from our diet and from other sources within our body. Yes, your mom was right — eat your fruits and vegetables. Our modern Western diet is low in fruit and vegetable consumption. This reduction of bicarbonate in our diet decreases our body's ability to neutralize acid production, leads to a low grade metabolic acidosis in our body, reduces connexin function, and thereby contributes to disease.

MACRONUTRIENTS-PROTEIN

Breakdown of protein (especially animal protein) in the diet leads to the generation of sulfuric acid. Would you allow anyone to administer an intravenous solution of sulfuric acid to you? By consuming high quantities of animal protein we are doing this to ourselves.

Mediterranean man, unlike Neanderthal, had the sense not to consume too much animal protein in his diet, keeping his consumption to less than 40% of total calories. Unfortunately, explorers to the "New World" (America) in their desperation for food failed to eat a balanced diet. They consumed primarily animal protein in their diet, and developed "rabbit starvation". This is what the American explorers called the excess protein condition that results in nausea, diarrhea, and even death. Ironically, today man embraces the Atkin's diet: a high protein, acid producing diet. In contrast, our ancestors ate lots of fruit and vegetables — good sources of dietary bicarbonate, necessary for balancing pH.

FATTY ACID COMPOSITION

Arachidonic acid is a fatty acid derived primarily from a diet rich in animal fat. This molecule is important to life and makes up 5-15% of all fatty acids in cell membranes. At the same time, arachidonic acid is the precursor of most inflammatory molecules and is thought to be involved in cancer, allergic and / or inflammatory conditions such as asthma and arthritis, and other inflammatory disorders. Reduction of this molecule in animal models has shown promise for reducing asthma, arthritis and kidney inflammation.

Arachidonic acid is a connexin closer. It has been shown to acidify the inside of cells, and this is considered the mechanism by which it closes connexins. It is increased by excess animal meat consumption, which is also linked to many of the diseases of modern man.

We do need arachidonic acid for life. But we also need to watch our intake of meat, so that our total consumption is kept low.

GLYCEMIC LOAD

Glycemic index is a relative comparison of the blood glucose (blood sugar) raising potential of foods. Processed (refined) grain and sugar products almost always have a higher glycemic index than do unrefined natural plant foods. For example, the glycemic index of processed bran flakes cereal is 74, while natural bran has a glycemic index of 42. High glycemic index foods promote insulin resistance, which in turn leads to elevated glucose levels in our body and type-2 diabetes. Mass consumption of all these high glycemic index foods is causing diabetes to become one of our nation's top diseases. Diabetes is growing at a rate soon to surpass cancer and heart disease.

Elevated glucose is detrimental to connexins — it closes them. Disruption of connexin 36 by glucose contributes to a loss of control of insulin-producing cells in the pancreas. Clinically, this disruption of connexins by elevated glucose may contribute to the changes in small blood vessels (microvasculature), peripheral nerves, and the lens of the eye that are seen in diabetic patients, and resulting in diabetic retinopathy (diabetes induced retina damage), peripheral neuropathy (diabetes induced nerve damage), and cataracts, respectively.

As we have seen in this chapter, the modern diet is predisposed to reduce the function of connexins. Now you know what types of food can do this. Think of how many times each day you consume foods that are known to diminish connexin function. Occasional indulgence is probably of no consequence. However, many of us regularly consume a diet that has an overall negative effect upon connexin function. The result is America leads the world in illness and in healthcare expenditure. Fortunately, good news is on the way — Chapter 5: Connexin Openers / Function Enhancers.

Let's Summarize

- What we eat can influence the function of our connexin molecules.
- The modern diet has evolved so rapidly that the human genome is unable to adapt rapidly enough to accommodate to the change.
- Today, we eat foods that acidify our bodies and reduce our connexin function.
- Man originally refrained from drinking seawater due to the salt. Now we not only consume too much NaCl (sea salt) but additionally consume salt substitutes containing chloride.
- Foods with a high glycemic load (high glucose) and a Neanderthal diet (high animal protein) are detrimental to our health and connexin function.

Connexins and the Food We Eat — II 5

"To keep the body in good health is a duty, otherwise we shall not be able to keep our mind strong and clear."

Gautama the Budda, 563 BC

"The physician is nature's assistant."

Galen

CONNEXIN OPENERS / FUNCTION ENHANCERS → HEALTH / LIFE

In the previous chapter, you learned about connexin closers or connexin function reducers and how the eight major diseases are linked to the connexin molecule and its function. There are molecules in our diet that open or enhance connexin function that will be discussed. If we eat the proper foods, our connexin molecules remain open and connected and operate so that we are healthy. Foods in our diet that enhance connexin function will be presented with respect to 3 of the fundamental nutritional characteristics of the ancestral human diet as compared to our current American diet: 1) acid-base (pH) balance, 2) micronutrient composition, and 3) fatty acid composition (see Figure 5.1).

FIGURE 5.1 Connexin Openers / Function Enhancers

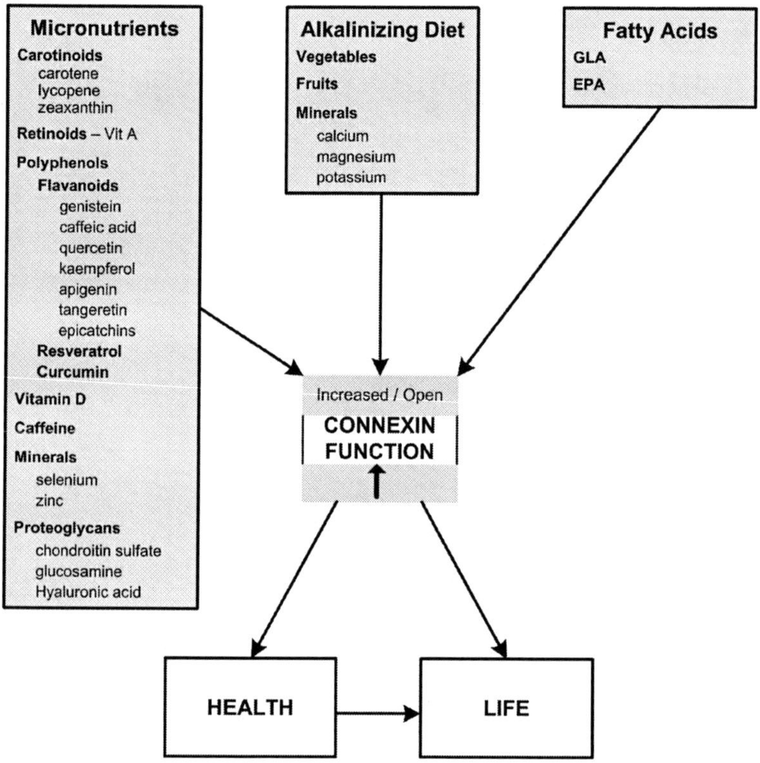

ACID / BASE BALANCE — ALKALI GENERATING FOODS

Medical and scientific studies show that the net acid producing diet of modern man along with declining kidney function (this reduces the ability to remove acids from our body) as we age, results in the formation of a metabolic state known as chronic low-grade metabolic acidosis. Too much acid in our body adversely affects all of our cells. As we become more acidic over time our connexin function may continually decrease, leading to disease.

The solution: reduce the acid generating foods in our diet and increase alkalinizing foods in our diet.

This concept was first proposed in the studies published by Sherman and Sinclair in 1907. Since that time, scientists around the world have refined these measures and determined the ability of different foods to produce or neutralize acid in our bodies. Remer and Manz developed a method to quantify net acid generation and excretion called PRAL (potential renal acid load), as discussed earlier.

This method is simple to use and represents acid generating foods as positive (+), and alkali or acid neutralizing foods as (-). For four of the major food categories the PRAL values are shown in Table 5.1; they are placed in descending order, from most to least alkalinizing. The portion size of each food is the same and equals 100 grams or about 3.5 ounces. In Chapter 8, individual food PRAL values will be given along with instruction on how to evaluate the net acidifying or alkalinizing effect of your diet.

TABLE 5.1 Major food groups that are net alkalinizing average PRAL per 3.5 ounces (100 grams)

Food Group	PRAL
Fruits and Juices	-4.44
Vegetables	-2.78
Beverages	-1.23
All Fats and Oils	0

As presented in Chapter 4, connexin function is enhanced by an alkaline pH environment.

In addition to opening and closing the connexin molecule, pH also regulates connexin gene expression (the creation of the connexin protein from the genetic code), such that an alkaline pH can lead to a greater number of connexins molecules per cell, and overall greater connexin functionality.

Prior to agriculture and the domestication of animals, man's original diet was usually net alkalinizing because of the greater availability of alkalinizing foods in central latitudes inhabited by most humans. A study by Sebastian et. al. showed that 117 of 137 diet scenarios for pre-agricultural man resulted in net alkalinization. A good example is the highly alkaline diet of one of our close relatives, the gorilla. His average urine pH is 8.45. Our now extinct predecessor, Neanderthal man, once consumed an acidifying diet. Now modern man has "advanced" his diet over the last several generations to be ever more acidifying. Without change, are we at risk of extinction?

MINERALS

Minerals such as calcium, magnesium, and potassium can have a beneficial influence on acid production and excretion in humans. These minerals lead to a reduction in net acid generation which produces a more alkaline urine pH. The net result of consuming these minerals, typically found in fruits and vegetables, is less systemic metabolic acidosis putting you in a net alkaline state which enhances connexin function. All three of these minerals are associated with improvements in osteoporosis and heart disease, which are both known to be associated with decreased connexin function. In addition, reduction of urine acidity prevents kidney stone formation and reduces existing stones. This reduction of stone formation can prevent loss of kidney function and restore reduced kidney function due to stone formation. Maintaining good kidney function helps to excrete acids from our body, making us less acidic overall.

Zinc is also a mineral which can improve connexin function. When levels of zinc are low in humans there is increased osteoporosis and cancer.

Selenium has been shown to reduce the toxic effects of a known connexin inhibitor, peroxynitrate; thereby, in effect, enhancing connexin function.

Minerals have traditionally been available to man either in water (natural spring waters) or in fruits and vegetables, as a byproduct of the soil they were exposed to. Unfortunately, our modern environment has left us deplete of minerals because our plant foods are now grown in mineral depleted soils and our water has been filtered and "purified". Centuries of agriculture and our processed irrigation and drinking water creates this deficiency of minerals. It has been shown that 60 to 73 percent of Americans over the age of 2 years are not meeting the US recommended dietary allowances for magnesium, calcium, potassium and zinc.

MICRONUTRIENTS

Micronutrients are nutrients that are present in foods in small quantities such as vitamins and phytonutrients. In this section I identify micronutrients that are connexin enhancers or openers. You may know some of these nutrients as being good for health, but you may not know the extent to which they improve health by enhancing connexin function.

CAROTINOIDS

Carotinoids are the yellow, orange and red pigments found in plants. The North American diet contains alpha-carotene, beta-carotene, beta-cryptoxanthin, lutein, zeaxanthin, and lycopene. Epidemiological and laboratory studies have shown that consumption of these molecules is associated with reduced risk of cardiovascular diseases and some cancers. All of these carotinoids molecules have been shown to enhance connexin number and / or function.

RETINOIDS

Vitamin A is the general term for many of the retinoid compounds, more specifically known as retinol, retinal, and retinoic acid. Alpha-carotene, beta-carotene, and beta-cryptoxanthin are provitamin A carotinoids, meaning they can be converted into vitamin A by our body. Vitamin A is important for vision, immunity, growth, and red blood cell production. And at the same time, Vitamin A enhances our connexin function.

POLYPHENOLS

Polyphenols (Figure 5.2) are one of the most numerous and ubiquitous groups of plant molecules and are an essential part of the human diet. These molecules are associated with many health benefits. Within the polyphenols group you find flavanoid molecules as well as resveratrol and curcumin. These are all connexin openers and enhance connexin function and number.

FIGURE 5.2 Polyphenols

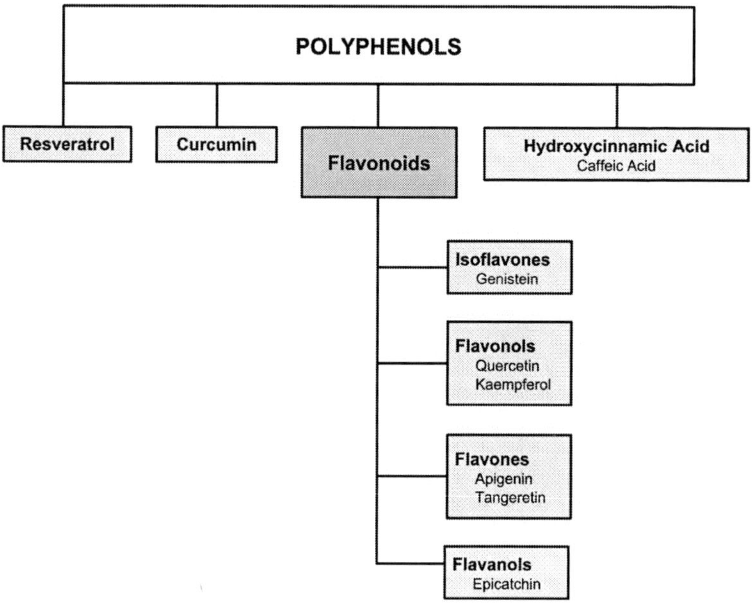

FLAVONOIDS

Flavonoids are a large family of polyphenolic compounds synthesized by plants. Many of these molecules are connexin enhancers including: genistein (soy), caffeic acid (coffee), quercetin (apple), kaempferol (capers), apigenin (parsley), tangeretin (tangerines), and epicatchins (tea). In many laboratory, animal and human studies these molecules have been shown to reduce cardiovascular disease, cancer, neurodegenerative disease, and many chronic degenerative diseases.

RESVERATROL

Resveratrol is best known as the beneficial molecule in red wine and is also found in grapes, grape juice and some berries. Resveratrol has been shown to be beneficial for reducing many of the processes associated with cancer and heart disease. This molecule is known to improve and enhance connexin function.

CURCUMIN

Curcumin is the major component of the spice turmeric. It is the yellow pigment that has been used in medicines in India and Southeast Asia for generations. According to tradition curcumin is useful to treat inflammation, skin wounds, liver disorders, and certain tumors. Modern medicine is currently evaluating this molecule for its anti-tumor properties and has shown it to be beneficial for reducing colon and breast cancers. This polyphenol molecule also enhances connexin function.

VITAMIN D

Vitamin D enhances connexin function and has been shown to prevent osteoporosis, cancer, autoimmune disease, and hypertension. Vitamin D is a fat-soluble vitamin essential for maintaining normal calcium metabolism. This molecule is part of our cholesterol synthesis pathway and its active form is created when the skin is exposed to ultraviolet-B radiation, which comes from the sun. Along with the sun, we get vitamin D from certain fish, like herring, mackerel and in cod liver oil.

CAFFEINE

Well known as a molecule in coffee and tea, caffeine has been used for centuries as a stimulant. The healthy benefits or detriments of the molecule have been debated for years with no clear answer. Caffeine makes us more alert and can reduce our sleep. This effect is consistent with its ability to open connexin molecules and enhance their function.

GLYCOSAMINOGLYCANS AND PROTEOGYLYCANS

These molecules are best known to improve arthritis and joint health and that is why they are used in many supplements. Chondroitin sulfate, glucosamine, and hyaluronic acid, all members of this group, have been shown to enhance connexin function.

FATTY ACIDS

Omega-3 fatty acids are polyunsaturated fatty acids, and include eicopentaenoic acid (EPA), which is commonly obtained through the consumption of fatty fish: salmon, mackerel, herring, trout and tuna. EPA has been shown to be associated with prevention and reduction in heart disease, stroke, cancer, diabetes, arthritis, and other inflammatory diseases. EPA is a known connexin opener and enhancer. Gamma-linolenic acid (GLA) is another fatty acid that is a connexin opener and has also been shown to be beneficial for skin disorders (eczema, psoriasis), arthritis, and cancer.

THE COMMON THREAD

All of these micronutrients have been investigated and shown to be beneficial in maintaining one's health and working to prevent most of the eight major diseases of modern man. The reasons proposed for the efficacy of these micronutrients have been many and diverse, but none have previously unified their beneficial actions as being attributable to enhancement of connexin function.

This common thread may not seem so astonishing when we consider that connexins are integral to the existence of all multicellular life forms, including ourselves. Beginning with our development as an embryo, all of our organs, our entire body for that matter, are formed by the presence and differential expression of connexins. As our connexins are starved for essential connexin enhancing micronutrients, our cellular continuity and communication are disrupted. We are in essence "falling apart". It makes perfect sense that our organs, tissues, and cells will no longer work together and we are more susceptible to disease and to dying prematurely!

Let's Summarize

- Our modern diet places most people in a state of "low grade metabolic acidosis".
- Foods that make us more alkaline can enhance our connexin function.
- Phytonutrients rich in polyphenols found in the "Mediterranean Diet" are beneficial to our connexins.
- We are deplete in connexin-enhancing minerals such as calcium, magnesium, potassium, zinc and selenium.
- Never before have we had the luxury of being able to so easily access as many foods and supplements that enhance connexin function.

Connexins and the Mind-Body Connection 6

> "Lack of activity destroys the good condition of every human being, while movement and methodical physical exercise save it and preserve it."
>
> Plato

> "If you ask what is the single most important key to longevity, I would have to say it is avoiding worry, stress, and tension. And if you didn't ask me, I'd still have to say it."
>
> George Burns

EXERCISE, STRESS, AND CONNEXIONS

From the earliest writings of man, exercise has been extolled as beneficial to both body and mind. More than 2,000 years ago Plato and Hippocrates advised that a lack of physical activity was detrimental to health. Systematic research into the effects of occupation upon health began when an Italian physician, Bernardini Ramazzini, looked at the diseases of different tradesmen from 1690 to 1731. He found that runners and professional messengers avoided the health problems of sedentary tailors. With the advent of The Industrial Revolution in

England in the mid 1800's, many sedentary occupations were created, and shortly thereafter it was observed that physically active workers had lower mortality rates than their sedentary counterparts. As industrialization grew in England and America so did disease and mortality. In 1912, JB Herrick described the rapidly increasing incidence of heart attacks. In 1939, Hedley, in the *US Weekly Public Health Report* looked at 5,116 deaths due to heart attacks and found higher mortality for white collar workers than laborers. The eight major diseases have continued to increase along with industrialization. Unfortunately, many physicians and medical scientists did not acknowledge or support such findings until late in the 20th century.

Exercise is good for you. Regular physical exercise has been shown to produce many health benefits. These include prevention of cardiovascular disease, cancer, diabetes, osteoporosis, and arthritis; and improvements in overall body fitness, psychological well-being (reduced depression and anxiety), life expectancy, and quality of life. A recent study associating exercise with health was the Harvard Alumni Health Study. A total of 13,485 men were studied. Vigorous activities, distance walked and fights of stairs climbed, independently predicted longevity. Active women also live longer, as validated by the Framingham Heart Study for women aged 50 to 74. Physical activity associated with longevity appears to be a universal finding. A prospective Swedish study of both men and women has shown that walking speed and physical activity are most strongly associated with performance in later life, age 66-76. In 2005, the Jerusalem Longitudinal Study correlated longer life with regular exercise, positive self-assessed health, activities of daily living, and independence.

Exercise has also been shown to reduce mental stress. It is known that physical fitness can insulate an individual from the potentially harmful effects of mental stress. Regular exercise has been demonstrated to reduce depression and anxiety in many studies. Even some measures of mental function have been associated with physical activity in a limited number of studies. These realizations are not new. The health benefits of combining physical exercise and mental stress reduction were well known to many ancient civilizations. They have been made popular today by the introduction of Eastern exercise forms such as Yoga and Tai Chi.

Many of these exercise and stress-reduction health benefits can be linked to improved connexin function. Let's start by looking at how exercise improves connexin function. Figure 6.1 shows that vigorous physical activity significantly increases four molecules in our body: growth hormone, ACTH (adrenocorticotropic hormone), cortisol and prolactin. All four of these molecules have been shown to enhance connexin function.

FIGURE 6.1 Physical activity and Connexin function

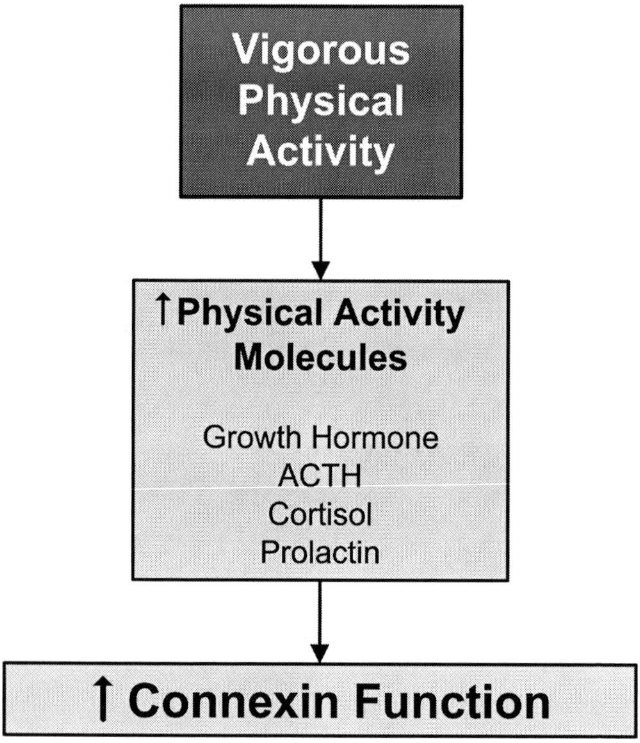

These four molecules peak at different times during the course of physical activity. Growth hormone and ACTH peak during maximal exercise (100%) to levels 12 and 4 times their resting baseline levels, respectively. Their elevated levels then rapidly return to resting levels within an hour. Prolactin increases to three times the resting level, five minutes following maximal exercise. Prolactin then returns to resting levels 1 hour after exercise has stopped. Cortisol increases five minutes following maximal exercise and then remains elevated during the entire initial conditioning and training phases which may last for weeks.

More importantly, this exercise-induced increase in molecules is exercise-intensity or workload related. For example, growth hormone increases in a workload dependent fashion. The growth hormone concentration in the blood is on average 7.4 mIU / liter, at rest; 15 mIU / liter at 40% maximal effort; 27.5 mIU / liter at 60%; 60 mIU / liter at 80%; and 87.1 mIU / liter at 100%. In contrast, ACTH, prolactin, and cortisol do not increase until one has reached 80% of maximal exercise effort.

This exercise-intensity dependent rise in these molecules is especially significant in light of the fact that studies do not show cardiovascular benefit from exercise until a vigorous level of activity (fast walking or running >4 mph, lap swimming, and cycling) is done two or more times a week, preferably for 45 minutes or more. Heart attack rates decrease from 5.2% to 2.4% in people of Body Mass Index (BMI) < 23; whereas, a reduction from 8.8 to 0.7% is seen in people with BMI's > 28. Recreational work such as gardening is of no benefit for preventing coronary heart disease.

Exercise studies show that until one achieves vigorous levels of exercise, one does not achieve significant cardiovascular benefit and that one's body does not produce significant increases in connexin-enhancing molecules.

An easy method for achieving 80% maximal effort is to exercise at 80% of your maximum heart rate. The easiest formula to use, which is reasonably accurate, is:

(220 - age) x 0.8 = 80% maximum heart rate

Age	80% Max Heart Rate
30	152
40	144
50	136
60	128
70	120
80	112
90	104

The most accurate method is a treadmill stress test administered by a professional. If you are over 35, overweight, have been sedentary for several years, or have a history of heart disease in your family, clinical testing is recommended.

Vigorous physical activity will enhance your connexin function and your health.

Mental stress, depression and anxiety abound in our society today. People talk about the Mind-Body Connection, but what does this mean and how does it occur? I believe that the connexin molecule serves as the keystone for this interaction, as it does for the eight major diseases. First, it is important to understand that mental stress results in our body producing numerous stress molecules. Figure 6.2 outlines how mental stress leads to both depression and anxiety, and in turn or concurrently produces mental-stress molecules (angiotensin II, endothelin I and mitogens). These three molecules have been shown to reduce connexin expression and function. It should come as no surprise that stress is also associated with the eight major diseases of modern man.

FIGURE 6.2 Mental stress

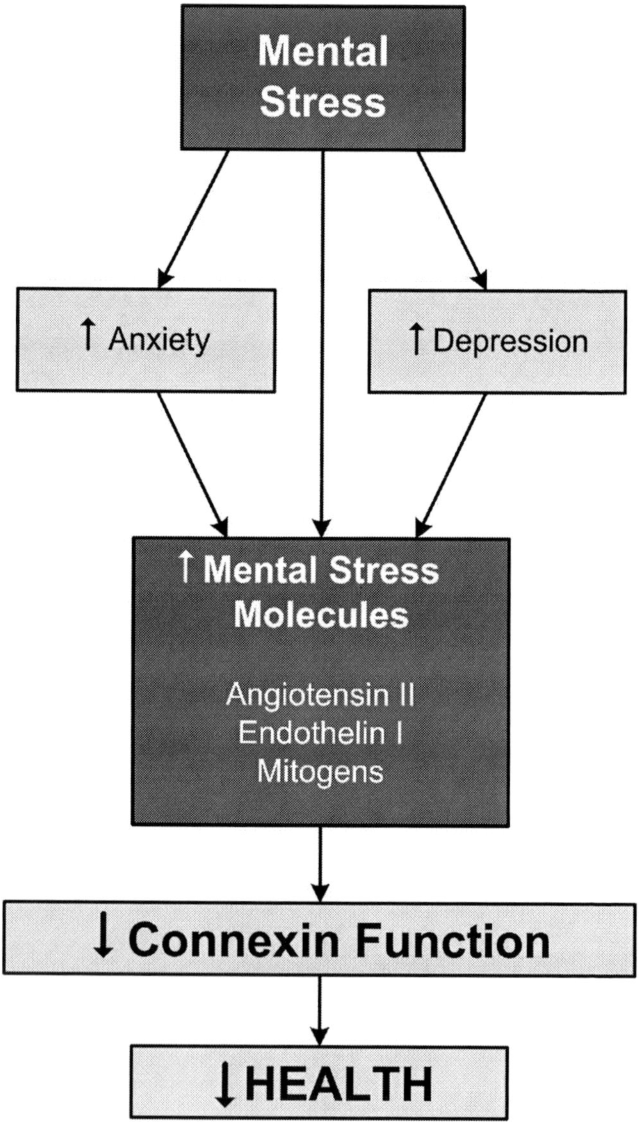

As Plato and Hippocrates recognized, exercise was beneficial for both the body and the mind. Exercise, is especially beneficial to connexin function because it increases both physical activity associated molecules (growth hormone, ACTH, cortisol and prolactin) while simultaneously reducing mental stress associated molecules. These two effects complement each other and enhance connexin function as outlined in Figure 6.3. Now it becomes even more apparent why exercise is so beneficial to health. **Vigorous physical activity will enhance connexin function and simultaneously reduce mental stress.**

FIGURE 6.3 Exercise

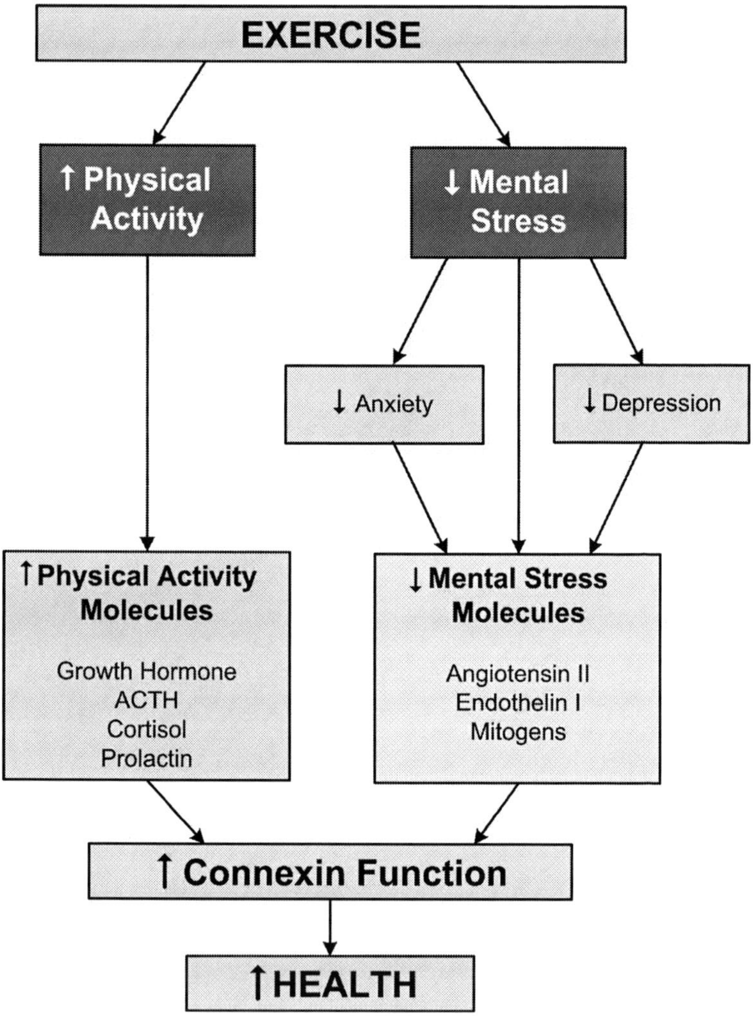

Figure 6.4 demonstrates how increased connexin function leads to improved health. Increased connexin function leads to decreased connexin-related disease (eight diseases of modern man), increased mental focus, increased muscle mass, increased bone density, and increased wound healing. These 5 elements then enhance physical and mental fitness which in turn improves the above 5 elements even more. This is what is called a positive feed-forward cycle which continues over time leading to optimal physical and mental fitness.

FIGURE 6.4 Increased Connexin function and improved health

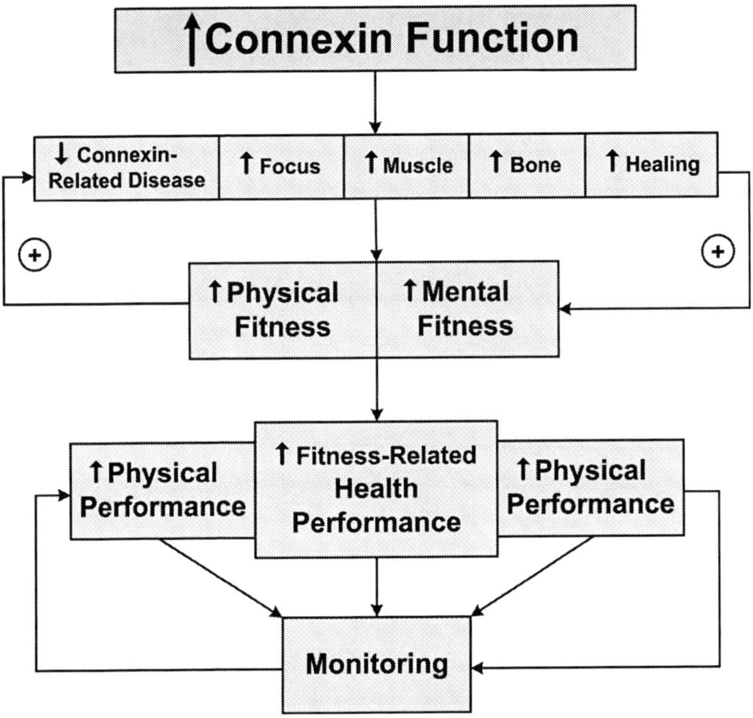

Enhancing one's physical and mental fitness then serves as the foundation to enhance performance, which I define as a measure of functioning or operating. Performance can be objectively measured as previously discussed. We can be high, average, or low performers. Examples of performance measurements are blood pressure, heart rate, mile running or swimming times, physical reaction times, depression or anxiety scores, measures of cognitive mental function (e.g. IQ), memory tests and numerous other measurements. By self monitoring our physical, mental and health performance we can improve overall or selected performance even more. This performance/monitoring cycle also is a positive feed-forward cycle. Together, enhanced fitness and enhanced performance lead to improved health and wellness.

The mind-body connection is real, and at the heart of it is the connexin molecule. A multitude of molecules, genes, billions of years of evolution, along with some divine intervention have all played essential roles; but at the crossroad sits the connexin molecule.

Let's Summarize

- Exercise is good for you.
- Vigorous exercise naturally stimulates the production of connexin-enhancing molecules.
- Exercise reduces mental stress leading to reduction of anxiety, depression and stress-induced connexin impairing molecules.
- Increased physical activity and reduction of mental stress promote connexin function.
- Overall, increased physical fitness and mental fitness enhance our daily performance.

So let's get started 7

"Let nothing which can be treated by diet be treated by other means."

Maimoindes

"To eat is necessary, but to eat intelligently is an art."

La Rochefoucauld

Phase I — Food and pH

In chapter 5 you learned about connexin closers or function reducers and in Chapter 6 about openers or function enhancers. Now in Phase I, all your new found pH knowledge will be put to practical use. You will learn to create pH balanced and alkalinizing meals. This is one of the major steps toward achieving the Connexin Molecular Fitness Lifestyle. In Phase II (chapter 8) you will learn about connexin enhancing phytonutrients. In Phase III (chapter 10) you will be introduced to connexins and exercise, and learn to implement the other important connexin enhancing components of The Connexin Molecular Fitness Lifestyle.

Let's get started!

We start correcting pH first because it involves a single easy measurement, urine pH, which can be quantitatively evaluated by you. This method allows you to learn about the effect of your diet upon your pH. You can keep records of your measurements and trend them; thereby becoming experienced with the value of record keeping and feedback (fully presented in Chapter 11) for making adjustments to optimize your diet. Before moving onto Phases II and III, I recommend becoming comfortable with Phase I.

Table 7.1 presents the major food groups and their average potential for either acidifying or alkalinizing your body. Recall that PRAL stands for potential renal acid load. A negative value

TABLE 7.1 Major food groups
Average PRAL per 100 grams/3.5 Ounces

Food Group	PRAL 100g (3.5 oz)
Fruits	-4.44
Vegetables	-2.78
Beverages	-1.23
Fats and Oils	0
Milk and non-cheese dairy	0.95
Bread, Grains	5.73
Eggs	8.21
Fowl and Fish	8.45
Meat	8.38
Meat Processed	11.42
Cheese Soft	4.51
Cheese Hard	19.22

indicates that the food item is alkalinizing. A positive value means a food is acidifying.

If you do not find a specific food item listed in Tables 7.2-7.9, refer back to Table 7.1 and use the PRAL value of the general group as a measure for that unlisted individual food item.

To help you in a more practical sense, Tables 7.2-7.8 list PRAL values for individual food items, which you can use for planning and creating daily meals. **The goal is to consume a net alkalinizing diet each day.** Take the time to familiarize yourself with these values. Then, we will move on to create a net alkaline diet. Notice that all fruits (7.2), nearly all vegetables (7.3) and most beverages (7.4) are alkalinizing.

TABLE 7.2 Fruits and fruit juices

Alkaline	PRAL 100g (3.5 oz)
Raisins	-21.0
Black Currents	-6.5
Banana	-5.5
Apricot	-4.8
Kiwi	-4.1
Cherries	-3.6
Pears	-2.9
Orange (Juice)	-2.9
Tomato (Juice)	-2.8
Pineapple	-2.7
Lemon (Juice)	-2.5
Apple (Juice)	-2.2
Strawberry	-2.2
Watermelon	-1.9
Grape (Juice)	-1.0

TABLE 7.3 Vegetables

Alkaline	PRAL 100g (3.5 oz)	Acid	PRAL 100g (3.5 oz)
Spinach	-14	Peas	1.2
Celery	-5.2	Lentils	3.5
Carrots	-4.9		
Zucchini	-4.6		
Cauliflower	-4.0		
Potatoes	-4.0		
Radish, red	-3.7		
Eggplant	-3.4		
Beans, green	-3.1		
Tomato	-3.1		
Lettuce	-2.5		
Chicory	-2.0		
Leeks	-1.8		
Onions	-1.5		
Peppers	-1.4		
Mushrooms	-1.4		
Broccoli	-1.2		
Cucumber	-0.8		
Asparagus	-0.4		

TABLE 7.4 Beverages

Alkaline	PRAL 100g (3.5 oz)	Acid	PRAL 100g (3.5 oz)
Orange Juice	-2.9	Most Soft Drinks	0.4
Tomato Juice	-2.8	Whole Milk Pasteurized	0.7
Lemon Juice	-2.5		
Red Wine	-2.4		
Apple Juice	-2.2		
Mineral Water (Apollinaris)	-1.9		
Mineral Water (US Brands)	< -0.2		
Coffee	-1.4		
White Wine	-1.2		
Cocoa	-0.4		
Tea	-0.3		
Beer	-0.1		

You can see in Table 7.5 that all oils are pH neutral, PRAL=0. Most nuts are acidic, but that is OK because you do not consume large quantities of them and many, such as walnuts, are very high in omega 3 fatty acids. Note that hazelnuts are alkalinizing. Nuts are generally beneficial for preventing heart disease.

TABLE 7.5 Oils and nuts

Alkaline	PRAL 100g (3.5 oz)	Acid	PRAL 100g (3.5 oz)
Hazelnuts	-2.8	Butter	0.6
Margarine	-0.5	Walnuts	6.8
Olive Oil	0	Peanuts	8.3
Sunflower Oil	0		

Table 7.6 shows that the majority of grain, flour, and starch based foods are acidifying. Note: the potato is alkalinizing.

TABLE 7.6 Bread, grain, and starches

Alkaline	PRAL 100g	Acid	PRAL 100g (3.5 oz)
Potatoes	-4.0	Rice, white, easy	1.7
		Wheat bread, whole	1.8
		Rye crisp	3.3
		Wheat bread, white	3.7
		Wheat bread, mixed	3.8
		Rye bread, mixed	4.0
		Rye bread	4.1
		Rye flour	5.9
		Corn flakes	6.0
		Noodles, egg	6.4
		Spaghetti	6.5
		Wheat flour	6.9
		Spaghetti, whole	7.3
		White flour	8.2
		Oat flakes	10.7
		Rice, brown	12.5

Tables 7.7, 7.8, and 7.9 identify that all high protein content foods are acidifying. These include: red meats, fowl, eggs and

fish. Notice that processed meat products result in a greater acidifying potential than do unprocessed meats. A recent medical study in the *Journal of the American Medical Association* shows that processed meats are associated with a higher incidence of contracting colon cancer than are unprocessed meats or fish.

TABLE 7.7 Fowl and fish

	Acid PRAL 100g (3.5 oz)
Haddock	7.1
Herring	7.1
Cod	7.1
Chicken	8.7
Turkey	9.9
Trout	10.8

TABLE 7.8 Meat and eggs

	Acid PRAL 100g (3.5 oz)
Eggs, white	1.1
Meat, lean	7.8
Pork, lean	7.9
Eggs, whole	8.2
Steak, lean and fat	8.8
Veal	9.0
Luncheon meat	10.2
Liver sausage	10.6
Salami	11.6
Corned Beef, canned	13.2
Eggs, yolk	23.4

The following table, 7.9, shows us that all minimally processed dairy products such as milk, yogurt, cream, and ice cream are slightly acidifying. Whereas, highly processed, hard cheeses are some of the most acidifying foods that exist.

TABLE 7.9 Milk products and cheese

	Acid PRAL 100g (3.5 oz)
Buttermilk	0.5
Ice cream, vanilla	0.6
Milk, whole, pasteurized	0.7
Yogurt, whole milk, fruit	1.2
Cream, sour	1.2
Yogurt, whole milk, plain	1.5
Soft cheese	4.3
Cottage cheese	8.7
Camembert cheese	14.6
Gouda cheese	18.6
Hard cheese	19.2
Cheddar cheese	26.4
Processed cheese	28.7
Parmesan cheese	34.2

Now that you are armed with the acidifying and alkalinizing potentials of the foods we eat, it is time to put everything we have learned together. You can create a net alkaline meal. Having studied the tables, you may think that you can only eat fruits and vegetables, but not so. **You can eat everything, in the proper balance**, and achieve alkalinization of your body. The Connexin Molecular Fitness Lifestyle is not a vegetarian program, unless you want it to be. I personally enjoy a balanced meal as long as it is net alkalinizing because of the variety of flavors, tastes and aromas it affords.

Let's start with breakfast.

Breakfast A	**PRAL**	**Breakfast B**	**PRAL**
Orange juice, 8 oz	-6.63	Milk, 8 oz	1.6
Coffee, 8 oz	-3.2	Coffee, 8 oz	-3.2
Tomato, 3.5 oz	-3.1	Wheat toast, 2 slices	1.8
Potato, 3.5 oz	-4.0	Corn beef, 3.5 oz	13.2
2 Eggs	8.2	2 Eggs	8.2
Total	**-8.73**	**Total**	**21.6**

Look at these two breakfasts. Breakfast A is net alkaline, despite having 2 acidic eggs. This was done by complementing the eggs with the proper selection of alkaline beverages and vegetables. On the other hand, Breakfast B is net acidic. It has the same number of food items, but more acid producing choices. As you can see, it is all about making the most-healthy choices. You are in charge! You are responsible. Easy isn't it?

Now that you are a pro, let's try dinner.

Dinner A	PRAL	Dinner B	PRAL
Mineral water, 12 oz	-6.5	Tap water	0
Red wine, 7.3 oz	-4.8	Beer, 12 oz	-0.34
Potato, 7 oz	-8.0	Pasta, 7 oz	14.6
Steak (lean) 7 oz	15.6	Steak (lean) 7 oz	15.6
Spinach, 3.5 oz	-14.0	Broccoli, 3.5 oz	-1.2
Total	-17.7	**Total**	**28.66**

Once again notice that Dinner A is alkaline, while Dinner B is acidic. Yet, both meals include the same 7 ounce steak.

These two meal examples demonstrate that one can eat acidic entrees such as steak and eggs and yet remain net alkaline by choosing alkaline side dishes and beverages. You're in charge and it is easy! Just plan ahead and you can have wonderful alkaline meals three times a day whether you are a meat-eater or a vegetarian.

How do you know this is working? Using the pH strips that are part of your monitoring program you can measure your urine pH and see whether your renal acid load is net acidic or alkaline. Most American adults, eating a typical American diet have an acidic urine pH of 5.5 to 6.5. Before you begin changing your diet establish your baseline urine pH by measuring it twice a day, daily, for 1 week and then average your values. As you will learn in Chapter 11, chart and graph your results.

Begin eating net alkaline meals and again measure your pH twice daily and obtain your new average. A goal to strive for is 7.0 - 7.5. Remember, pH is measured on a log scale, and that means if you change the pH value by 1 point, you have actually produced a 10-fold change in hydrogen ion concentration. Compare the results and your trends to see for yourself what a difference different food choices make.

A net alkaline diet makes your body less acidic and helps maintain the naturally alkaline state of your blood and cells. Along with alkalinizing our body come an enhancement in connexin function and a reduction in disease. This is your first step toward optimal health.

The next chapter is Phase II — Connexin Enhancing Phytonutrients. You will learn about the different types and amounts of connexin enhancing molecules found in various foods and learn to optimize your meals using this information as well.

Let's Summarize

- PRAL stands for potential renal acid load.
- A negative PRAL value indicates that a food item is alkalinizing.
- A positive PRAL value means that a food is acidifying.
- A net alkaline diet makes your body less acidic, enhances connexin function and reduces disease.

IT'S NEVER TOO LATE 8

"Leave your drugs in the chemist's pot if you can heal the patient with food."

Hippocrates

"Man's mind is his basic tool of survival. His mind is given to him, its content is not. His body is given to him, its sustenance is not. He cannot obtain his food without a knowledge of food and of the way to obtain it."

Ayn Rand, *Atlas Shrugged, 1957*

PHASE II — Connexin Enhancing Phytonutrients and Micronutrients

Phytonutrients are nutrients found in many vegetables and fruits and to a lesser extent in grains and seeds. These molecules are not vitamins or minerals, but are biologically active constituents of many pigments found in your food. Chapter 5 introduced you to phytonutrient and micronutrient connexin enhancers. In chapter 8 these molecules will become foods. This chapter contains tables of each connexin-enhancing phytonutrient and micronutrient and the foods in which they are found in the greatest amounts.

Now that you have mastered and optimized your pH in Phase I (chapter 7), you are ready to embrace Phase II and enhance your connexin function even further by including foods that contain these connexin-enhancing phytonutrients to your diet. Chapter 9, Connexin Cuisine, will put all of these connexin-enhancing foods together into wonderful recipes and meals.

Carotinoids

Carotinoids are the yellow, orange and red pigments found in plants. These colors are a great aid to identifying which fruits and vegetables contain large amounts of these pigments. The carotinoids have been shown to be beneficial for cancer, cardiovascular disease, age-related macular degeneration and cataracts. Table 8.1 lists the foods which contain the highest amounts of beta-carotene, lycopene and zeaxanthin per 3.5 ounces or 100g.

TABLE 8.1 Carotinoids

Molecule	Food Item	ug (microgram) / 100g (3.5 oz)
Beta-Carotene	Carrots	18,250
	Grape leaves	16,194
	Kale	9,226
	Sweet Potato	9,180
	Pumpkin	6,940
	Spinach	5,597
	Swiss Chard	3,954
	Beet Greens	3,405
	Cilantro	3,440
	Collards	3,323
	Apricots	2,554
	Red Pepper	2,379
	Cantaloupe	1,595
	Broccoli	779
Lycopene	Tomato, paste	29,330
	Tomato, sauce	15,916
	Tomato, soup	10,920
	Tomatoes, canned	9,708
	Watermelon	4,868
	Tomatoes, raw	3,025
Zeaxanthin	Corn, canned	528
	Persimmons, raw	488
	Spinach, raw	331
	Cornmeal	457
	Turnip Greens	267
	Collards, cooked	266
	Lettuce	187
	Kale, cooked	173

RETINOIDS

Vitamin A is the general term for many of the retinoid compounds. Some of the carotinoids listed above (alpha-carotene, beta-carotene) can be converted by our body to vitamin A and are called provitamin-A carotinoids. This connexin enhancer supports normal reproductive function, immune function, gene expression, and vision. Studies have also documented reduced cancer. The vitamin A content of the provitamin-A foods can be expressed as retinol activity equivalents (ug RAE). The recommended dietary allowance (RDA) for vitamin A was revised by the Food and Nutrition Board (FNB) of the Institute of Medicine in 2001. The RDA for vitamin A is given as ug/day or RAE. The adult Male amount is 900 RAE and female 700 RAE. Eating foods with too much preformed vitamin A, not carotinoids, can cause toxicity when long-term consumption of vitamin A is more than 3000 RAE or 10,000IU per day of preformed vitamin A.

TABLE 8.2 Retinoids

Molecule	Food Item	IU/(3.5 oz)	RAE/(3.5 oz)
Vitamin A	Turkey, giblets, cooked	35,790	10,737
	Beef, liver, cooked	26,088	7,744
	Sweet Potato, baked	19,218	961
	Kale, cooked	14,704	681
	Carrots, raw	12,036	602
	Spinach, cooked	9,377	524
	Lettuce, raw	7,405	370
	Peppers, red, sweet	3,623	157
	Melon, cantaloupe	3,381	168
	Papaya, raw	1,108	55

Polyphenols

All of the polyphenols presented in this chapter are connexin enhancers. These molecules are associated with many health benefits. Higher intakes of polyphenols are associated with reductions in disease. For example (Table 8.3), caffeic acid, a hydroxycinnamic acid, part of the polyphenol group commonly found in coffee, is a connexin enhancer.

FIGURE 8.1 Polyphenols

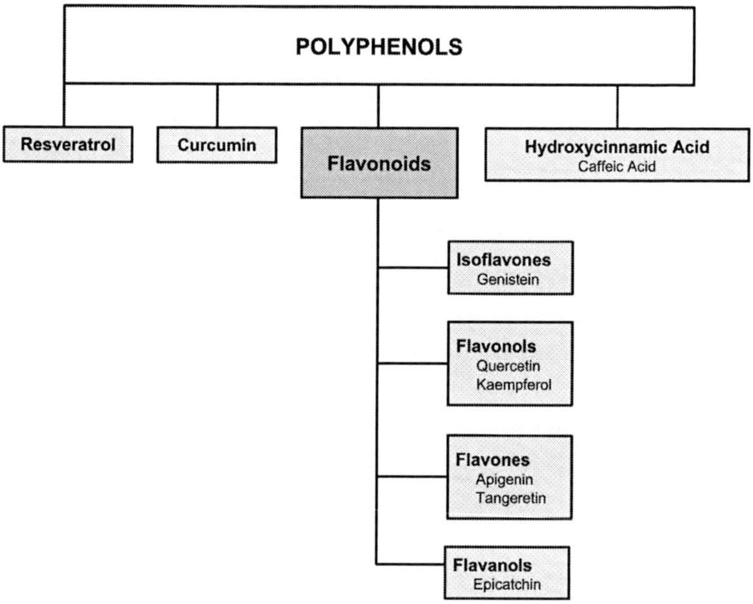

TABLE 8.3 Hydroxycinnamic Acid

Molecule	Food Item	ug (microgram) / 100g (3.5 oz)
Caffeic Acid	Coffee	96
	Carrot	14
	Potato	3.6
	Apple	2.4
	Tomato	2.4
	Red Wine	1.8

Flavonoids

Flavonoids are a large family of polyphenolic compounds synthesized by plants. Many of these molecules are connexin enhancers. These molecules have been shown to reduce cardiovascular disease, cancer, neurodegenerative disease, and many chronic diseases in laboratory, animal and human studies. No adverse effects have been associated with high intakes of flavonoids from plant-based foods.

TABLE 8.4 Isoflavones

Molecule	Food Item	mg / 100g (3.5 oz)
Genistein	Soy Flour	98
	Soybean	60
	Natto	29
	Miso	25
	Tofu	21
	Soy Milk	6.1

TABLE 8.5 Flavonols

Molecule	Food Item	mg / 100g (3.5 oz)
Quercetin	Capers	180
	Fennel	48.8
	Elderberry	46
	Peppers	15 - 30

TABLE 8.5 Flavonols (continued)

Molecule	Food Item	mg / 100g (3.5 oz)
Quercetin	Onion	20
	Cranberries	14
	Kale	7.71
	Apple	6
	Black Currents	6
	Blueberries	4
Kaempferol	Capers	135
	Kale	27
	Watercress	13
	Chives	10
	Fennel	6.5
	Broccoli	6

TABLE 8.6 Flavones

Molecule	Food Item	mg / 100g (3.5 oz)
Apigenin	Parsley	302
	Celery Hearts	19
	Peppermint	8.7
	Thyme	5.0
Tangeretin	Tangerine, juice	4.1
	Tangerine, peel, extract	1,310
	Commercial Orange Juice	0.26mg
	Orange Oil	2.6

TABLE 8.7 Flavanols

Molecule	Food Item	mg / 100g (3.5 oz)
Epicatchin	Green Tea	20 - 80
	Dark Chocolate	41.5
	Broad Beans	22
	Milk Chocolate	10.5
	Black Tea	10
	Apples	8
	Apricots	6

RESVERATROL

Resveratrol, a known connexin enhancer, is the beneficial molecule in red wine. Resveratrol has been shown to be beneficial for reducing many of the processes associated with cancer, heart disease and to increase the lifespan of certain forms of life. Resveratrol is not known to be toxic or cause adverse effects in humans.

TABLE 8.8 Resveratrol

Molecule	Food Item	mg / 100g (3.5 oz)
Resveratrol	Red Wine	3 - 4
	Grape Juice	1 - 9
	Grapes	1.5 - 7.5
	Blueberries	0.7 - 5
	Peanuts	0.7 - 5

Curcumin

Curcumin, another connexin enhancer, is the major component of the spice turmeric; it is the yellow pigment that has been used in medicines in India and Southeast Asia for generations. Modern medicine is currently evaluating this molecule for its anti-tumor properties and has been shown to be beneficial for reducing colon and breast cancers.

TABLE 8.9 Curcumin

Molecule	Food Item	mg / 100g (3.5 oz)
Curcumin	Turmeric	500 - 5,000

Vitamin D

Vitamin D enhances connexin function and has been shown to prevent osteoporosis, cancer, autoimmune disease, and hypertension. Vitamin D toxicity is very unlikely in healthy people at intake levels lower than 10,000 IU/day.

TABLE 8.10 Vitamin D

Molecule	Food Item	IU / 100g (3.5 oz)
Vitamin D	Cod Liver Oil	10,000
	Herring	1,628
	Mackerel	360
	Sardines, canned	272
	Caviar	232

Fatty Acids

EPA has been shown to be associated with prevention and reduction in heart disease, stroke, cancer, diabetes, arthritis, and other inflammatory diseases. EPA is a known connexin opener and enhancer. Gamma-linolenic acid (GLA) is another fatty acid that is a connexin opener and has also been shown to be beneficial for skin disorders (eczema, psoriasis), arthritis, and cancer. Practically, the most abundant palatable and cost effective source of GLA is hemp oil. Below in Table 8.11 and 8.12 are the foods most abundant in these fatty acids.

TABLE 8.11 EPA in Raw Fish

Molecule	Food Item	grams / 100g (3.5 oz)
EPA	Mackerel, wild, Atlantic	0.898
	Herring, wild Atlantic	0.709
	Salmon, farmed, Atlantic	0.618
	Salmon, wild, Atlantic	0.321
	Tuna, wild, Bluefin	0.283
	Trout, farm	0.260
	Trout, wild	0.167
	Sea Bass, wild	0.161
	Swordfish, wild	0.108
	Halibut, wild	0.071
	Catfish, farmed	0.067
	Haddock, wild	0.059
	Cod, wild	0.010

TABLE 8.12 GLA

Molecule	Food Item	grams / 100g (3.5 oz)
GLA	Borage Oil	25
	Black Currant Oil	14
	Evening Primrose Oil	7
	Hemp Oil	3

This chapter, combined with the previous one, provides you with all the information necessary to optimize your connexin function by eating the right foods. You can see that many of these connexin enhancing molecules are touted for their health benefits in general. Now, you understand why these phytonutrients and micronutrients are so essential to your health and well being — they enhance the function of your connexin molecules. The tables provide you guidance on selecting the foods which contain the highest amounts of a given connexin enhancing molecule. Using these food items in your regular diet you will be able to create wonderful and delicious meals. To help you understand how to optimize your meal preparation, read Chapter 9, Connexin Cuisine, replete with instructions and recipes for well balanced and easy to prepare connexin-healthy meals.

Let's Summarize

- Optimize your connexin function by eating the right foods.
- Phytonutrients found in many fruits and vegetables are essential to your health.
- Micronutrients like the fatty acid EPA found in salmon are essential to your health.
- Use the food charts in this chapter to select foods with high amounts of connexin-enhancing molecules.

Connexin Cuisine 9

"One can not think well, love well, or sleep well if one has not dined well."

Virginia Woolf, 1929

"Our lives are not in the lap of the gods, but in the lap of our cooks"

Lin Yutang

We have learned how our diet can influence our connexin function. Now we can create healthy meals that can improve our health and wellness through enhanced connexin function.

The recipes to follow are examples of the wonderful healthy meals and diet you can enjoy by following *The Connexin Molecular Fitness Lifestyle*. These meals will show you that healthy food does not need to be less than delicious. All of these recipes have been connexin-optimized by incorporating connexin-enhancing nutrients and alkalinizing foods into them.

The focus of this food is health not weight loss. Although, eating a connexin-healthy diet should help you lose weight. If you consume too much food (calories) for your level of activity, then weight gain will occur. There is nothing specifically "bad" about naturally occurring fats, proteins, or carbohydrates that have been part of our diet over millennia, they are all good for our body's functions and many are essential for life. Our meals will optimize the natural nutrients that provided us advantages over other species, millions of years ago. A selection of healthy food items is provided below to familiarize you with the great meals you can have and create using connexin-enhancing foods.

SNACKS AND APPETIZERS

Let's Be Nuts

- 1/3 pound Hazelnuts
- 1/3 pound Walnuts
- 1/3 pound Raisins

Mix all three ingredients together and eat as a connexin friendly snack.

Tantalizing Tapenade

- 1/4 cup whole almonds shelled
- 1/4 cup whole hazelnuts shelled
- 1/2 cup pitted green olives
- 2 cups parsley with stems discarded
- Grape seed oil added until preferred consistency achieved
- 1/2 cup sliced bottled jalapeno peppers
- 3 tbsp fresh Lemon juice
- Fresh cracked pepper

Toast almonds and hazelnuts in an oven at 325 degrees until lightly toasted, do not burn. Next, blanch the parsley. Then add the parsley, olives, almonds and hazelnuts to a food processor on pulse mode. Add grape seed oil in small amounts to the mixture in the food processor until a smooth, but not creamy consistency is achieved. Add lemon juice and pepper to the mixture for taste. Chill in refrigerator and serve as a spread on crackers or approximately 1/4 inch thick French baguette slices.

Garbanzo Carotinoids

4 medium carrots, sliced
2 garlic cloves crushed
1 yellow onion
– chopped
6 ounce can of chick peas
drained and rinsed
1 ounce of pecans
crushed
1 tsp coriander
1 tsp ground ginger
1 tsp white pepper

Cook the carrots until tender. Sauté garlic, onion, pecans, pepper, ginger and coriander in 2 tsp avocado or hazelnut oil. Once sautéed place mixture into food processor and blend with carrots and chick peas until smooth. Chill in refrigerator and serve on crackers or French baguette slices.

Salmon Deviled Eggs

4 Hard-boiled eggs cut
in halve lengthwise
2 tsp soy mayonnaise
2 tbsp light sour cream
⅓ cup finely diced
smoked salmon
3 tsp minced red onion
2 tsp capers
Fresh ground black
pepper to taste
Chives for garnish

Carefully remove egg yolks and place in bowl and mash them with a fork. Add mayonnaise and sour cream, mixing until smooth. Then add the salmon, fennel, onions and capers and pepper to taste. Fill the egg whites and garnish with chives.

BREAKFAST RECIPES

A Red, White, and Blue Breakfast

- ½ cup fresh or frozen blueberries
- 1 cup plain Soy milk
- ½ banana
- 6 whole strawberries
- +/- ½ peach

Place all ingredients into a blender and blend until frothy.

Serves 1 or 2

Currant Affairs in the Morning

- ⅓ cup plain cream of wheat
- ⅔ cup Soy milk
- 1 tsp unrefined sugar
- ½ banana
- chopped walnuts
- 2-3 heaping tbsp of black currants

Place ingredients into a sauce pan (and place on the stove) or Pyrex measuring cup (place into the microwave for about 3 minutes) and cook to personal texture.

Serves 1 or 2

Connexin Concoction – A Juice to Start the Day

2 large organic carrots
1 organic Fuji apple
½ bunch parsley
¼ fennel bulb
1 piece peeled ginger, 1 inch long
6 oz fresh squeezed orange juice

Serves 1 or 2

To make the juice use a screw type juicer capable of extracting juice from thin leafy vegetables such as parsley or wheat grass. This makes 2 servings of refreshing juice.

Ou-la-la Omelette

2 whole organic eggs
½ onion – chopped
¼ bunch parsley – stems removed
¼ Bulb fennel – chopped
½ red bell pepper – chopped
½ tomato – diced
¼ cup baby spinach
¼ cup feta cheese
Salt and fresh ground black pepper
2 tbsp hazelnut oil

Serves 1 or 2

Combine bell pepper, tomato, fennel, onion, and parsley into skillet with hazelnut oil and sauté until tender, and drain off any water. Next add salt, pepper, spinach and beaten eggs, cook to preferred consistency and sprinkle feta cheese on top and then fold.

SALADS

Tangerine Spinach Salad

- 3-4 Tangerines – sections separated, save any juice
- 6 ounces stemmed baby spinach leaves
- 2 heads baby endive, leaves separated
- 1 fennel bulb thinly sliced into rings
- 1 red onion, sliced into thin rings
- 1 bunch watercress, large stems removed
- Fresh ground pepper and salt to taste
- 1 clove thinly sliced garlic
- 4 tbsp walnut oil
- 2 tbsp shelled hempseed

Serves 4 to 6

Prepare all items, place in bowel and toss. Mix with premixed walnut oil, salt, pepper, tangerine juice and garlic. Arrange items to maximize visual appeal and sprinkle with hempseeds.

Blueberry, Fig, Avocado Spinach Salad

- 6 ounces stemmed baby spinach leaves
- ⅓ cup fresh blueberries
- 1 fresh ripe avocado cut into 8 sections
- 6 small ripe fresh figs cut into ¼ inch rings
- 2 tbsp Cider vinegar
- 4 tbsp grape seed oil
- 1 tsp Ground ginger
- 1 tbsp Dijon mustard
- ½ lemons worth of juice
- 2 tbsp shelled hempseed

Serves 4 to 6

Mix dressing out of vinegar, grape seed oil, lemon juice, ginger and mustard. Toss salad ingredients with salad dressing. Sprinkle with shelled hempseeds and serve chilled.

Field Greens with Beets, Tangerines, and Feta Cheese

6-8 ounces mixed gourmet field greens
½ - ¾ pound balsamic beets
1 cup fresh tangerine sections
¼ cup crumbled feta cheese
1 tbsp lemon juice
3 tbsp extra-virgin olive oil
1 tbsp balsamic vinegar
Fresh ground pepper
Grated tangerine rind
2 tbsp shelled hempseed

Serves 4 to 6

Make dressing by mixing balsamic vinegar, olive oil, lemon juice, pepper, and tangerine rind. Toss salad ingredients with salad dressing; then add sectioned beats, tangerine sections and feta cheese and sprinkle with shelled hempseeds.

Apple, Fennel, and Walnut Salad

1 fennel bulb (about 1 pound)
1 pound Fuji apples
½ cup toasted walnut halves
3 tbsp minced fresh parsley
2 tbsp raspberry vinegar
3 tbsp walnut oil

Serves 4 to 6

Trim fennel bulb, quarter and thinly slice. Peel and core apples, quarter and thinly slice. Toss apples and fennel with oil and vinegar, salt and pepper to taste. Add walnuts and parsley on top.

ENTREES

SUCCULENT SNAPPER

2 pounds Red Snapper fillet – skin removed
2 yellow onions chopped
8 garlic cloves – crushed in garlic press
2 tbsp olive oil
¼ cup white wine
2 limes juiced
4 large tomatoes chopped or 28 ounce can of chopped organic, low salt, tomatoes
20 large green pitted olives stuffed with jalapenos, sliced
¼ cup chapped fresh destemmed parsley
1 tbsp rosemary
1 tsp oregano
2 bay leaves
2 tbsp capers
Freshly ground pepper
Thin lemon slices
1-2 cups cooked Basmati rice

Serves 4

Rinse the filets and set in a bowl immersed in lime juice and set in refrigerator. Use a large skillet (14 inch diameter or larger) with a cover. First sauté the onions in olive oil and wine, covered, until tender and then add the garlic. Place the skillet on low heat and add the tomatoes, sliced olives, parsley, rosemary, oregano, bay leaves, and pepper. Cover and simmer for 30 - 45 minutes. Uncover, place on medium heat and add the capers and snapper fillets and lime juice to the skillet, immersing the fillets and cook until fillets tender.

Serve the fillets onto a bed of cooked white rice and cover with the sauce and garnish with lemon slices.

Primo Pasta Pronto

- 4 cups al dente cooked angle-hair pasta
- 2 tsp olive oil
- 2 garlic pressed garlic cloves
- 1 tomato, diced
- 3 tsp capers
- 1 lemon zest
- 1 tsp oregano
- ½ cup crumbled feta cheese

Serves 4

Heat the olive oil in a skillet on medium heat and sauté the garlic and tomatoes for 2 minutes, then add in oregano, feta, capers, lemon zest for another 1 minute. Toss mixture with angle hair pasta and serve.

Savory Salmon

- 2 pounds Salmon fillet – with skin
- 1 lemon juiced
- 4 tbsp Dijon / horseradish mustard
- 2 tbsp capers

Serves 4 to 6

Rinse the filets and set on a plate skin side down, cover with lemon juice and then spread mustard mixture onto fillet. Next sprinkle capers onto mustard covered salmon. Place fillets skin side down onto medium heat grill, cover and cook until fillets are tender and mustard sauce has slightly darkened. Remove salmon from skin and serve.

Tangerine Chicken

- 4 – 6 pounds Whole "free range "roasting" chicken
- 6 tangerines – sectioned into slices
- 2 tsp grated tangerine rind
- ⅓ cup tangerine juice
- 8 pieces peeled sliced fresh ginger
- 3 tbsp grated fresh ginger
- 2 tbsp low-sodium soy sauce
- 2 tbsp honey
- 1 tbsp sesame oil
- 1 cup low-sodium chicken broth
- ¼ cup sherry
- 2 tbsp herbed bread crumbs
- 1 tbsp all-purpose flour

Serves 8

Place bread crumbs, grated tangerine rind, 2 tbsp tangerine juice, grated ginger, soy sauce, honey, and sesame oil into a bowl and set aside. Wash chicken and loosen skin over breast and drumsticks and insert above mixture. Place pieces of ginger 4 halves of tangerine sections into central chicken cavity. Place chicken breast side up into broiling pan and bake for 20 minutes at 450 degrees.

Next, reduce the oven temperature to 350 degrees and bake an additional 1 hour 30 minutes or until meat thermometer in thigh is at 180 degrees. Then, place chicken on platter and keep warm.

Combine 1 cup chicken broth with flour and cooked chicken drippings and whisk in sherry and tangerine juice. Bring to a boil and cook until thickened while stirring. Serve with chicken by ladling on.

Pork Loin with Figs, Apple, and Cider Sauce

1 boneless pork loin roast (4 pounds)
1 cup Pinot Gris
1 tbsp lemon juice
12 to 14 dried figs
3 cups apple cider
3 tbsp hazelnut oil
2 cups coarsely chopped onions
2 cups of cubed apples
3 garlic pressed cloves
1 tbsp of sage
Fresh ground pepper

Serves 6 to 8

Heat 3 tbsp of hazelnut oil in skillet and add onions and apples and cook on medium heat until golden, then add sage and peeper to taste. Turn off heat and let cool.

Make syrup and figs by adding 2 cups apple cider and figs together in a saucepan and cooking until liquid reduced to about ¼ cup. Turn off heat and separate figs and syrup. Cut up figs and add to mixture of apples and onions.

Pre-heat the oven to 350 degrees. Cut the pork loin lengthwise, about ¾ of the way through. Sprinkle cut opening with pepper and spoon in fig/apple/onion mixture. Tie pork with string, place into roasting pan and bake in oven occasionally turning until meat thermometer reaches 135 degrees, and then turn broiler on for a few minutes to crisp outside of pork roast. Take roast out of oven and let sit covered with foil to keep warm.

Make a sauce for the roast by combing 1 cup apple cider, wine, and browned scrapings from cooked pork roast pan. Boil mixture to reduce, then reduce to warm and add lemon juice and pepper to taste. Cut pork roast into slices, arrange on individual plates and cover with sauce. Garnish with any excess fig/onion/apple mixture and fresh sage leaves.

SIDE DISHES

WHEAT PILAF

3 tbsp extra virgin olive oil or hazelnut oil
½ cup thin egg noodles (or orzo)
1 cup coarse bulgur
2 cups low salt chicken stock

Serves 4

Cover the bottom of a saute pan with extra virgin olive oil or hazelnut oil and heat for 1-2 minutes. Add the egg noodles and stir until the noodles turn golden brown. Remove the saute pan from heat and let it cool.

In a second saute pan, bring the chicken stock to a boil. Add the bulgur, mix well and then turn down the heat to medium low. Stir in browned noodles with remaining oil and allow the bulgur and noodles to cook until the fluid is fully absorbed, about 15 minutes (note: if you use fine or medium bulgur, it will absorb faster). You are ready to serve!

Tangerine Yams

- 4 yams, well scrubbed and cut in half in the long direction
- 1 tbsp hazelnut oil
- ¾ cup pure maple syrup
- 1 cup tangerine juice
- ¼ cup organic brown sugar
- 1 tsp nutmeg
- 2 tangerines sectioned

Serves 4

Arrange the yam halves in a baking pan and pre-heat the oven to 350 degrees.

Mix the hazelnut oil, maple syrup, tangerine juice, brown sugar, and nutmeg in a bowl and ladle over yams in dish. Bake until yams are tender while basting with sauce mixture in pan.

When done, add tangerine slices for a few minutes to yams and then they are ready to serve.

Currants and Cumin Rice

- 2 tbsp extra virgin olive oil or hazelnut oil
- 2 tsp cumin seed
- 1 cup basmati rice
- 2 tbsp dried currants
- 2 cups chicken broth

Serves 4

Combine the oil, cumin seed, rice and currants in the saucepan and stir together on low heat until fully coated and fragrant. Add the chicken broth and bring to a boil. Cook covered, over low heat until the water is absorbed.

Spinach Orzo

1 cup Orzo
⅓ cup olive oil
3 tbsp fresh lemon juice
½ tsp lemon zest
2 cups chopped baby spinach
¼ cup red onion chopped
Cherry tomatoes for garnish

Serves 4

Boil orzo in 2 cups water until done and let cool. Sauté the red onion in olive oil until tender and then place the skillet on low heat. Add lemon juice and zest to sautéed onion and then add cut up pieces of baby spinach to mixture in skillet and wilt spinach slightly. Turn off heat and add cooked orzo and mix. Garnish with halved tomatoes around spinach orzo on plate.

Crispy Potatoes

6 medium red potatoes cut into quarters
1 tbsp hazelnut oil
Salt to taste
Fresh black pepper to taste
2 tsp oregano

Serves 4

Pre-heat oven to 425 degrees.

Mix potatoes with oil, salt, pepper, and oregano in bowl and when full coated place potatoes in pre-oiled baking dish for about 30-45 minutes stirring occasionally until golden. Then remove from oven and serve.

Gingered Carrots

1 tbsp avocado oil
1 pound small – medium carrots, peeled
¾ cup water
A 2 inch piece of fresh ginger peeled and sliced thin
¼ cup finely chopped parsley
2 tbsp organic cane sugar

Serves 4

Heat a skillet with hazelnut oil and add ginger and sauté for 1-2 minutes. Add carrots and water to pan, cover, and cook on low heat until carrots tender. Add sugar and cook turning carrots until liquid reduced and carrots glazed, then add parsley and coat carrots and serve.

Baked Fennel

4 pounds fennel bulbs
½ cup Olive oil
¼ cup freshly grated parmesan cheese
Freshly ground black pepper

Serves 4 to 6

Trim fennel bulbs by cutting off the stalks, base and outer leaves, and then quarter bulbs and remove any hard regions.
Blanch the quarters in boiling water for 8-10 minutes until tender. Place fennel into a shallow baking dish, drizzle evenly with olive oil, sprinkle with fresh ground pepper and parmesan cheese, and then place into a 400 degree pre-heated oven to bake about 10 minutes or until golden.

DESSERTS

Carrot and Walnut Cake

A

3 cups All-purpose flour
2 tsp baking soda
2 tsp baking powder
½ tsp salt
4 eggs
½ cup hazelnut oil
½ cup unsalted butter
2 cups organic cane sugar

B

2 cups shredded carrots
1 cup chopped walnuts
½ cup raisins (red flame)
2 tsp ground cinnamon

In a large bowl mix group **A** ingredients together until smooth. Then fold in group **B** ingredients uniformly. Pour mixture into buttered baking pan.

Bake for 70 to 75 minutes at 300 degrees, remove and cool on baking rack.

Chocolate Sorbet

1 cup natural cocoa powder
¾ cup granulated sugar
1 tsp vanilla
2 cups water
½ cup toasted hazelnuts (chopped)

Serves 4

Into a 4 quart sauce pan place cocoa powder and sugar. Add ½ cup water and mix to break up and lumps and then add ½ cup water and mix into a smooth paste. Add 1 cup water and simmer mixture on low heat for 3-5 minutes. Remove from heat and chill in refrigerator for 3 hours.

Place chilled mixture into ice cream maker and when mixture begins to thicken add the chopped toasted hazelnuts.

Apple Banana Crumble

2 bananas peeled and sliced
2 apples, peeled, cored and sliced
Juice of ½ fresh lemon
¼ cup organic brown sugar
¼ cup flour
¼ cup unsalted butter

Serves 4

Mix the banana and apple slices in a bowl with the lemon juice. Mix the sugar, flour and butter together until crumbly (topping mix). Layer the fruit into a glass baking casserole and top by sprinkling the topping mix over.

Place into 325 degree pre-heated over for 30-40 minutes.

Serve with vanilla ice cream.

Peach Blackberry Cobbler

- 2 cups fresh peaches peeled and sliced
- 1 cup fresh blackberries
- 2 tbsp grated lemon peel
- 2 tbsp melted unsalted butter
- 2 tsp baking powder
- ⅛ tsp salt
- 1 cup organic cane sugar
- ¾ cup milk
- 1 cup All-purpose flour

Serves 6

Pre-heat oven to 350 degrees.

Pour melted butter into glass ovenproof baking dish. Mix all non-fruit ingredients in bowl until smooth and pour into baking dish. Next add fruit and grated lemon peel evenly to batter in dish.

Bake for 45 minutes. Remove and serve warm. Adding vanilla ice cream is a nice touch.

Tropical Fruit Salad

- 2 Mangos, about 1 pound each
- 1 Papaya, 1 pound or larger
- 1 Pineapple, 3-4 pounds
- 2 large Bananas

Serves 6 to 8

Peel the mangos and cut the flesh off of the pit and place into bowl. Cut the papaya in half, scoop out the seeds, peel each half and cut into small pieces and place into bowl with mango pieces. Peel and core the pineapple and then cut the flesh into small pieces and add to the mango-papaya mixture along with any juice.

Place in refrigerator to chill. Before serving slice the bananas and add to chilled fruit mixture and serve.

These recipes are a few of the many healthy and delicious recipes you can create by using connexin healthy foods and molecules. Mix and match any of the recipes to make what you would like for lunch or dinner. Be adventurous and create your own recipes and meals by using the tables of connexin healthy molecules and foods in Chapter 8 and the alkalinizing food tables in Chapter 7. Enjoying great meals every day will keep you motivated to live *The Connexin Molecular Fitness Lifestyle.*

ConneXercise 10

"True enjoyment comes from activity of the mind and exercise of the body; the two are united."

Alexander von Humboldt (1769 – 1859)

"Fitness — If it came in a bottle, everybody would have a great body."

Cher

Phase III — Exercise and Relaxation

Exercise and relaxation are integral and complementary to proper nutrition. They are an essential component of your journey toward achieving molecular-fitness. In this chapter we introduce you to ConneXercise: an integrated health, nutrition and fitness program that optimizes physical and mental performance. ConneXercise is not just another diet, weight loss or stress reduction plan. It is an integrated health for life molecular-fitness program.

ConneXercise, Phase III, will complement and complete Phases I and II (chapters 7 and 8, respectively) by introducing exercise and stress reduction. Before you begin rigorous physical activity, you want to optimally prepare and build your body through diet and proper nutrition in order to maximize the benefits exercise will bring while simultaneously minimizing possible injury. **Phases I and II prepare you for Phase III, ConneXercise.**

Maintaining your proper weight combined with proper nutrition, exercise and stress reduction is healthy. Exercise alone is not a good weight loss intervention, but is a good method for preventing weight gain and maintaining one's weight. Exercising while obese can be dangerous. Our skeleton was not created to carry 10's, let alone 100's of extra pounds of weight. Exercising when obese is an invitation to destroy one's knees, hips, and ankles. Exercising when obese can destroy one's spine and may lead to a heart attack. Numerous studies (some referenced in this book) have shown that mild to moderate exercise alone does not enhance weight loss during dieting, nor does it lead to a prolonged increase in one's resting metabolic rate. Therefore, it is important to reduce one's excess weight prior to intense exercise.

You began this journey by alkalinizing your body in Phase I. One benefit of alkalinization was weight loss, if you needed it. An acidifying diet promotes weight gain, and makes fat breakdown next to impossible even when exercising. Now, provided your food intake does not increase, you should naturally maintain your weight and be able to lose weight if needed.

Chapter 7 (Phase I) and Chapter 8 (Phase II) of *The Connexin Connection* have enhanced your connexin function in preparation for ConneXercise. Your body is now prepared to optimally benefit from exercise and heal itself if injury were to occur. Before starting ConneXercise, establish your baseline health parameters and trends as presented in Chapter 11, such as your vital signs (blood pressure, heart rate, body mass index, glucose, pH) and Panel 20 Laboratory values, so that you are able to follow your progress and know when it is time to advance your program. Through ConneXercise you will maximize your fitness, optimize your performance, improve your health and extend your life.

Figure 9.1 illustrates the ConneXercise program — just follow along! ConneXercise begins with four elements: pH correction (alkalinization) — Phase I; connexin enhancing phytonutrients (micronutrients) — Phase II; exercise (cardiovascular / endurance, strength training, stretch / mobility); and relaxation or mental stress reduction. All of these connexin enhancing nutrients can be obtained from natural foods, but obtaining the quantities and entire selection through the food you eat can be very difficult. The addition of supplements can complement your connexin healthy diet and make it complete. These four elements all enhance connexin function by acting at the molecular level of your body's function.

Exercise needs to be introduced based upon your existing level of fitness and your health status. If you are not currently an exercise enthusiast or not actively followed by a physician, I recommend consulting a doctor prior to starting any exercise program.

Once you have been given the go ahead "clean bill of health," you need to select the types of exercise you like the most, from the ones which promote cardiovascular health the most (vigorous walking, running, bicycling, vigorous swimming, rowing, squash, cross country skiing). It is extremely important to select activities you enjoy because exercise studies have shown that you will stick with your program and stay motivated if you enjoy what you are doing. On average, I recommended 45 – 60 minutes of vigorous physical activity three days per week. In addition to your cardiovascular / endurance exercise, it is important to complement this with strength training, stretching, and relaxation training. Get ready to enjoy the experience.

FIGURE 9.1 ConneXercise

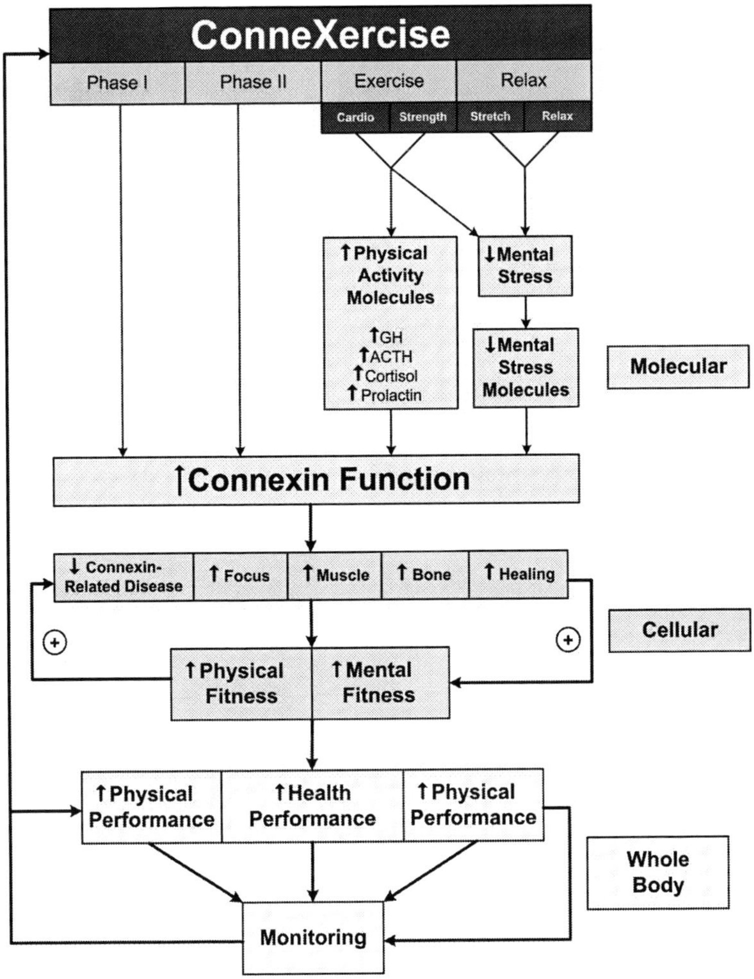

As we learned in Chapter 6, physical activity increases four molecules that directly enhance connexin function at a molecular level. Simultaneously, mental stress reduction also acts at the molecular level to decrease stress-related molecules that diminish connexin function. Together, these complementary effects are additive and even synergistic.

The increased connexin function obtained at the molecular level then leads to five cellular level improvements:

1. Prevention of the **Eight Major Diseases of Modern Man.**
2. Enhanced **Mental Focus.**
3. Increased **Muscle Mass.**
4. Increased **Bone Density.**
5. Enhanced **Wound Healing.**

These five cellular level physiologic improvements gained by following the Connexin Molecular Fitness Program enhance both physical and mental fitness which in turn feed-forward to further increase these five parameters. Once your cellular function is optimized, you will begin to see improvements in your performance; physical, health-related and mental. Now, you have improved your whole body's function.

Quantitatively measuring your whole body performance will enable you to monitor and chart your progress so that you can customize your ConneXercise program to optimize your health. As your connexin function improves (molecular level) your fitness will increase (cellular level) and your performance (whole body level) will improve as well. **Your body and mind**

will work better together. ConneXercise will take you to a new state of health and wellness. Don't get frustrated or discouraged. Remember, good things are on their way.

Let's Summarize

- ConneXercise is connexin enhancing nutrition and supplementation coupled with a comprehensive exercise and stress reduction program.
- A comprehensive exercise program includes strength or resistance training, cardiovascular workouts, stretching and relaxation.
- Increased connexin function at the molecular level leads to cellular improvement: the prevention of the eight major diseases, enhanced mental focus, increased muscle mass, increased bone density, and enhanced wound healing.

Who's monitoring those connexins? 11

"The patient should be made to understand that he or she must take charge of his own life. Don't take your body to the doctor as if he were a repair shop."

Quentin Regestein

"Each patient carries his own doctor inside him. They come to us not knowing that truth. We are at our best when we give the doctor who resides within each patient a chance to go to work."

Albert Schweitzer

TAKING CONTROL OF YOUR HEALTH: SELF-MONITORING AND YOUR MEDICAL RECORDS

Now that you have empowered yourself and taken charge of your health how do you monitor the status of your health — how it was, how it is, and what it may be in the future?

Today, it is possible to monitor your own vital health parameters and to chart and trend them. You can do the same with your doctors' or hospitals' diagnostic tests and medical records.

As part of *The Connexin Connection* program, I have created an easy to use and easily accessible web-based system for securely charting, tracking and storing your health information at HealthConnexin.com. Tracking one's health parameters and records have been invaluable in my medical practice over the last 20 years for diagnosing, treating and optimizing patients' health and wellbeing.

Yes, our bodies come pre-wired with the most advanced monitoring and control systems we can ever know, and some we will never comprehend. Without our knowledge or intervention, hundreds if not thousands of biological feedback control systems regulate our body, second to second, minute to minute, and day to day, making life possible. Our internal monitoring system is essential for life. Without this regulation, we would become ill and die. This is exactly what happens when any of these monitoring and feedback control systems fail. First, we become ill. Then, if the system remains dysfunctional and does not repair itself quickly, or we do not intervene and correct the malfunction, we soon die.

So what is a feedback control system? To illustrate, let's look at a system we are all familiar with and use: our home heating and cooling system. It took us thousands of years to design a system to control home temperature. Our bodies, however, monitor and regulate their own temperature at 98.6 degrees throughout our lives, in all climates, and during all forms of physical activity.

In order to keep our homes at a comfortable temperature we install a furnace to heat and an air conditioner to cool. Unregulated, either device runs the risk of disaster. If the furnace is uncontrolled it may become so hot that our house burns down. If the air conditioner is uncontrolled, it may become so cold that we can't live in our home. To prevent these disasters we install thermostats. Home temperature can then be set at a comfortable 72 degrees. Thermostats measure home temperature continuously. If the temperature is below 72 degrees the furnace is turned on. Then, when it is too hot, the furnace is turned off. If the temperature rises above 72 degrees, the air conditioner is turned on until the temperature is again below 72 degrees, then the air conditioner is turned off. All this temperature monitoring and control is done without our knowledge, 24 hours a day. This is known as a feedback control system.

We monitor many trends in our daily lives. Would you buy a stock based upon a single quote, at one point in time? Probably not! Information about the current and past history of the stock price would be important to know. Trend analysis, coupled with current market conditions, allows us to make an informed decision; whether to purchase the stock or wait and gather more information before proceeding. Similarly, we monitor the stock price and trend before selling.

Money is another good example of monitoring and trend analysis. Before credit cards, cash served as the medium for financial transactions. Cash allowed one to directly monitor and trend one's financial situation. Credit cards took away this immediate monitoring system. By eliminating feedback control, credit cards have allowed over-spending and have led to financial problems for many people.

If you own a business you hire an accountant to monitor and chart the financial health of the business. You use this information to make operating decisions and plans for the future of the business. You follow trends and use them to make predictions on the potential future growth and profitability of your company. These financial records are then used to value your company should you choose to sell it.

Monitoring is clearly integral to all forms of life. We monitor and trend most aspects of our personal lives, especially those we consider important.

What is more important than your health and well being? What is more important than to monitor oneself, to determine trends and to regulate one's own health?

"A wise man ought to realize that good health is his most valuable possession" — Hippocrates

If we have such wonderful feedback control systems built into our body why do we need to monitor ourselves?

We need to monitor ourselves because our lifestyles are generally not in accord with how we were created to live and often exceed the natural limits or capabilities of our biological feedback systems. We abuse or stress our bodies beyond what our natural control systems can regulate and repair, leading to disease and disrepair. We have been led to believe we can abuse our bodies in any fashion, then visit the doctor and get repaired. This is simply not true.

Many people circumvent financial monitoring by using credit cards and then go into debt and bankruptcy. In the same fashion many circumvent health monitoring then become ill and die prematurely.

The time has come to take responsibility for your own health and wellbeing. This begins with self-monitoring.

Do you allow your physician to make a medical treatment decision based upon a single measurement (blood pressure, heart rate, weight, glucose, or any single laboratory test value) taken during your annual physical ? It makes good sense to repeat, and if possible repeat again, vital signs and laboratory test measurements at different intervals before making a therapeutic decision. An important factor in medical decision-making is to have baseline data, as well historical data. This helps determine if there has been a change, either positive or negative, in a given measurement or if the measurement value is remaining stable. Also, measurements made under a person's daily environmental conditions are optimal. For example, if you have arrived late at the doctor's office after racing to avoid being late, then it is likely that your blood pressure and heart rate will be elevated. Just the fact that you are visiting the doctor's office may cause these measurements to be elevated. These one-time measurements should not be used as a basis to start you on medication to control your blood pressure.

WHERE ARE YOUR MEDICAL RECORDS?

In an ideal world, analyzing trends for vital signs or laboratory measurements is the best way for making therapeutic decisions in medicine. Most of the time, however, consecutive measurements over time are not available to the physician. Today's employment and medical systems make it very difficult to obtain historical data on a patient. How many times over the last 20 years, or even five years have you changed physicians? How many times over the last 20 years, or even five years have you gone to different physicians because your insurance plan changed or you moved? In today's world changing employment is the rule, rather than the exception; and along with changing employment comes changing health plans, physicians, and perhaps relocating your home. If you have remained with the same physician, it is still unlikely that she / he has a graphical record of your vital signs or laboratory test values available for quick and easy comparison. Do you have a copy of your medical records? Did you even know you could have them? Do you know how to retrieve your records? Do you recall all the doctors you have seen over time? For most of us, the answer is no. Yet, the fact remains that we are responsible for our medical information.

TAKE CHARGE OF YOUR MEDICAL RECORDS

In today's ever-changing medical environment it is more critical than ever for *you* to be in charge of your medical records. After all, who is more invested in your health than you and your family? The good news is that all of your medical records are available to you for the asking. Easy to use, reliable self-monitoring equipment and home laboratory testing kits are available for purchase by you at reasonably low prices in many retail and wholesale stores. These home tests can be just as, or more, accurate than those used in your physician's office or hospital. Blood pressure, heart rate, body mass index (weight), blood glucose, and urine pH are the easiest, most inexpensive and reliable tests to obtain. These vital measurements are great to record for both self-monitoring and to assist your doctor in regulating your health. These self-measurements are invaluable for early detection of medical conditions you and your physician may be unaware of. Early detection and awareness of a condition as a result of monitoring improves your chances of correcting a condition before permanent damage occurs and provides important feedback to your treating physician for optimizing your treatment regime.

Self-monitoring and diagnoses of medical conditions will play an ever greater role in your medical care. Secure records of your vital signs, laboratory tests, and other diagnostic studies conducted either by you or your physician are valuable assets in managing your own health. Remember, you are the person most invested in your health. Keeping track of and trending the laboratory tests results your doctor has ordered is very important. Standard tests that are usually ordered by your doctor include a Panal-20, complete blood count (CBC) and lipid profile. If you have other special testing needs you should obtain, chart and trend those results as well.

Over the years, do you know how you have been doing? Are you better, the same or worse this year than last year or 5 years ago? When told you are doing fine or are normal — is that good? Normal is usually interpreted as 50th percentile +/- one standard deviation, but are you a 34th or 99th percentile person? Are you better than average or worse? Without your test results and the data range you have no way of knowing where you stand and many times even your doctor doesn't know. Once you think about this, I am sure you will want to know.

WEIGHT — BODY MASS INDEX (BMI)

What does weight alone tell us about our health — more than you might expect. A simple calculation, Body Mass Index (BMI) = (weight in pounds / height in inches^2) x 703, adjusts body weight for height. According to The United States Center for Disease Control (CDC), BMI is considered one of the best methods for population assessment of overweight and obesity. BMI correlates with body fat. For example, a 160 pound person who is 6 feet tall has a BMI = [160 lbs / $(72)^2$ inches] x 703 = 21.7, which is the midpoint for the "normal" range. The BMI may not always accurately portray a person's weight status, especially for highly muscular individuals. The following table is provided by the United States Centers for Disease Control (CDC).

BMI	Weight Status
Below 18.5	Underweight
18.5 - 24.9	Normal
25.0 - 29.9	Overweight
30.0 and above	Obese

The BMI ranges are based on the effect body weight has on disease and death. Overweight and obese individuals (BMI of 25 or greater) are at increased risk for the following conditions:

- Diabetes
- High blood pressure hypertension

- High cholesterol
- Heart Disease
- Stroke
- Arthritis
- Some types of cancer
- Complications of pregnancy and poor reproductive health
- Psychological disorders
- Respiratory disorders and sleep apnea
- Gallstones
- Bladder control problems

BMI can be followed and used in conjunction along with other self-tests and your physician to diagnose and to treat potentially life-threatening medical conditions.

BLOOD PRESSURE

High blood pressure is a major risk factor for heart disease, stroke, renal disease and peripheral vascular disease. Currently, it is estimated that 28.6% (almost 1 in 3) of the U.S. population has hypertension (blood pressure greater than the "normal" range), while only 63.4% (2 out of 3) of those individuals are aware of their condition. Of those with hypertension, only 45.3% (less than half) are under treatment and in only 29.3% (less than 1 in 3) is their hypertension controlled by any means. In 2000, the US Department of Health and Human Services held a conference and set a goal for reducing the proportion of adults with hypertension from 28 to 16 percent by 2010.

A giant step toward reaching this goal can be made by home-blood-pressure-measurement. This provides individuals with the ability to detect hypertension and through feedback to one's physician, optimize individual treatment. At first glance the idea of self-measurement may seem inaccurate. Extensive research has been done by physicians and scientists throughout the world that prove self-measurement is more accurate and reliable than measurements done in the physician's office. In the physician's office "the white coat phenomenon" exists. Anxiety or an alerting reaction raises blood pressure measurement significantly.

Home blood pressure measurement has been shown to significantly improve long-term hypertension control, diagnose masked hypertension, improve patient compliance, and to assist physicians in improving the diagnosis rate for hypertension from 26 to 64%.

For hypertension related conditions home blood pressure measurement has helped diagnose hypertension in chronic hemodialysis (kidney failure) patients. It provides early detection of pre-eclampsia or high blood pressure in pregnant women. Home blood pressure measurement helps to control and prevent vascular (blood vessel) complications in type-2 diabetic patients, and has served as an independent predictor for stroke in the general population.

BLOOD GLUCOSE — DIABETES

Diabetes is a major health care problem in the United States. It is estimated that nearly one third of diabetes cases in the United States go undiagnosed. From 1997 through 2002, the number of new cases of diabetes increased from 878,000 to 1,291,000 (a 47% increase). The prevalence of diabetes has more than doubled from 5.8 million to 13.3 million from 1980 through 2002, with people 65 years or older comprising almost 40% of the diabetic population.

Self-monitoring of blood glucose levels has become commonplace for diabetic patients and has improved their glucose control. The new self-monitoring devices that measure blood glucose are simple to use and are relatively inexpensive. Nonetheless, nearly one third of diabetes cases in the United States remain undiagnosed. The self-monitoring of blood glucose levels can help to diagnose previously unrecognized diabetes. In addition, self-measurement of body mass index and blood pressure are both relevant and complementary to blood glucose measurement since 80.5% of diabetics are overweight, 46.1% obese and 63.3% have hypertension.

URINE pH

Urine pH is easy to measure. It is an inexpensive health parameter one can measure at home. Many studies have been done which demonstrate that urine pH is determined by one's diet and also reflects blood and intracellular pH, Figure 11.1.

FIGURE 11.1 Urine pH determination

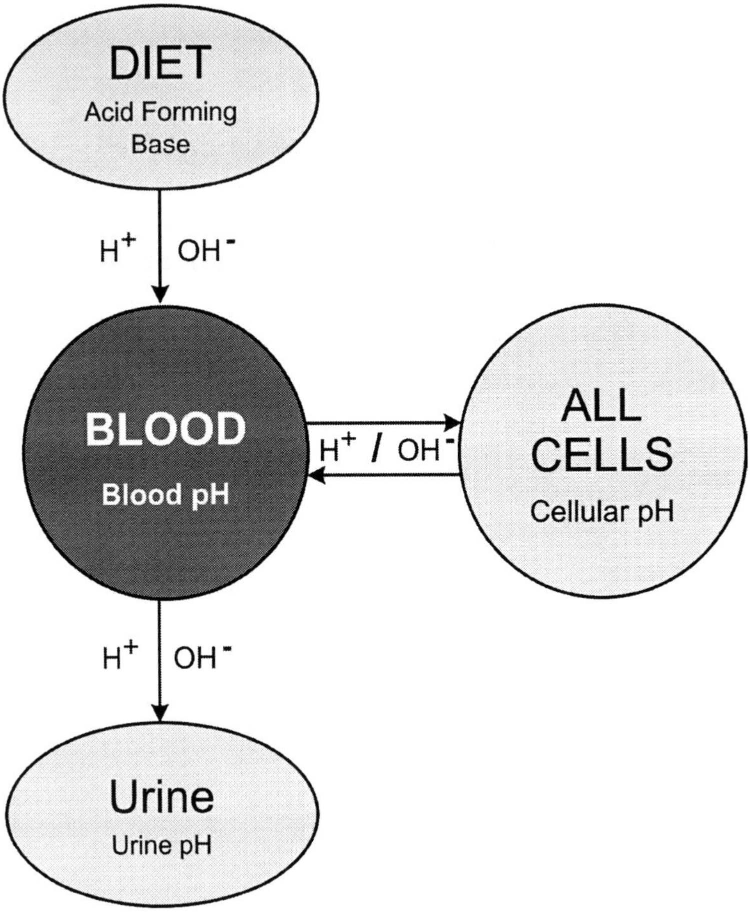

Urine pH is a measure of the body's production and excretion of acid. Highly acidic urine pH occurs when one eats foods that are highly acid generating. High acid production, leading to low grade metabolic acidosis, has been associated with numerous chronic medical conditions such as arthritis, cancer, Alzheimer's disease, heart disease, osteoporosis, ocular disease, and hypothyroidism. Therefore, the inexpensive and easy measurement of urine pH can be invaluable as a measure of how acidifying one's diet is, and thereby serve as an excellent self-monitoring device for improving and/or maintaining a healthy diet.

Remember, a single measurement at a given point in time is almost useless, unless it is so radically abnormal as to require immediate medical intervention. Monitoring is essential and serves as the foundation for your data trends. Analysis of these trends provides you with feedback on the direction your health is taking and may take in the future. Monitoring provides you useful and effective information on how well your lifestyle is working to maintain or improve your health.

Let's Summarize

- Our body is naturally designed to monitor itself each moment of every day. However, our modern lifestyle defeats many of our natural monitoring systems.
- Trends in our vital signs and laboratory test values are more informative than a single reading done at one point in time.
- Self-monitoring of many vital signs (BP, HR, blood glucose) can be superior to values obtained in medical settings.

12 Open and connected: the molecular fitness lifestyle

"Your children are not your children. They are the sons and daughters of life's longing, for itself."

Kahlil Gibran

"We have the power to make this the best generation of mankind in the history of the world — or to make it the last."

John F. Kennedy

You might say, why put all of your eggs in one basket, the connexin basket? The answer: the connexin basket has been scientifically proven to be valuable, it provides an explanation for all of the crucial nutritional characteristics of our ancestors. It encompasses the significantly known and scientifically validated nutritional findings and provides an underlying scientific rationale for their inclusion and integration. Simply put, the function of the connexin molecule is the single best determinant of our state of health and wellness.

There are and will always be thousands of "miracle" nutritional supplements, diets, and cures proposed. I find most lacking in scientific rationale. Can we afford to chase every mythical bird and mirage? Connexins and their role in our unified cellular structure are an irrefutable part of our human bodies and our well being. Healthy connexins make us whole and dysfunctional connexins are known to be linked to all of our major diseases. Yes, there are thousands of other important molecules in our body necessary for life, but that alone does not mean that they are equally important for health and life. The connexin molecule is the "keystone" of life. As an analogy, just think of the keystone in an ancient building. It was the central stone amongst many other stones that allowed the structure to support the archways. All of the other stones were important and necessary for the structure, but without the keystone the structure of the building failed, other stones came apart, and the building crumbled to the ground. Embracing the connexin lifestyle will build a solid foundation for building upon our body's "Molecular Keystone", the connexin molecule. Health and wellness, the ultimate freedoms, will be ours.

Health and wellness give one the ultimate peace of mind. With health comes freedom; the freedom to travel; freedom to enjoy and take care of yourself and your family; the freedom to earn a living; the freedom to stay healthy; freedom from healthcare services and expenditures; doctor's bills, hospital bills, nurses, prescriptions, medical insurance premiums, and nursing homes; freedom to plan for the future; freedom to enjoy every day!

Molecular Fitness is within your reach! As you have learned to this point, connexins are intimately involved in your life, but more critically they are involved in the prevention of disease and the extension of life. I divide your journey to molecular fitness into phases that are easy to accomplish and provide you with meaningful feedback as you progress.

Phase I is directed at pH, only one component of connexin enhancement, but a critical one. It is easily measurable on a real-time basis, providing nearly immediate feedback on your progress. Don't be in a hurry to move on. There is no rush. It is much better to learn in small steps because you will retain your newly acquired knowledge more thoroughly. Remember, if you are 30 years old that means you are 1,596 weeks in the making, and at 50 you have been under construction for 2,636 weeks, so don't expect changes to occur over night.

Once you have learned to regulate your diet and corrected and stabilized your pH, move on to Phase II — Connexin Enhancing Phytonutrients. As explained in Chapters 5 and 8, these molecules can be obtained in natural foods or via supplements.

The combination of correcting your pH and the use of connexin fortifying molecules prepares you for Phase III, optimizing physical and mental fitness. You probably have been engaging in physical exercise, but now your body is prepared as never before to benefit from these activities. Once you have achieved molecular fitness through Phase I and Phase II, you are ready for Phase III — a complete mind-body restoration and optimization program.

THE CONNEXIN LIFESTYLE AND YOUR FAMILY

Remember you are not only doing this for yourself, but also for your family. If you are planning a family, and I address this to mothers to be, what could be better than being personally healthy and fit so that you have an easy pregnancy and create a healthy child. As the old adage goes, "You are what you eat." And at no time is this more important than when you are planning a family. **Your child is not only what you have eaten, s / he even contains some of the molecules that your mother ate**.

Connexins are crucial to the entire developmental process leading to the creation of your child and in addition are integral to labor. What could be more important than optimizing not only your connexin function, but to also optimize your child's connexin function during development?

Once born your child's health and nutrition remain at your discretion. Knowing the importance of connexins to your child's development and health provides you with a safe natural formula to optimize your child's present and future performance. No need to stop what you have been doing for your child during pregnancy; allow your children to reach their full potential by optimizing their connexin function as they grow and mature.

Your parents and family on the other hand may be unaware of how to optimize their health and life, but it is never too late to benefit. The Connexin Lifestyle will provide benefits in the preventative and restorative direction at almost any age or condition. Partial restoration of function in those with chronic illness is akin to a miracle and will greatly enhance their quality of life. So don't hesitate, one of the greatest gifts you can give your family is to share *The Connexin Molecular Fitness Lifestyle* with them.

By practicing this lifestyle, you will achieve Molecular Fitness. It will become second nature. Your children will benefit even more, having been given a head start by you.

Enjoy your journey toward Molecular Fitness with *The Connexin Connection*!

Epilogue

"The causes and principals of different things are in a sense different, but in a sense, if one speaks universally and analogically, they are the same for all."

Aristotle – On Causes, In Metaphysics, Book XII

"To everything there is a season, and a time to every purpose under the heaven."

From Ecclesiastes

THE CYCLICAL NATURE OF DISCOVERY

Man has believed in the cyclic nature of discovery for thousands of years. Ideas come and go, only to be reborn. This cycle of thought is based upon insight or intuition, observation, experience and experimentation. The adherence to and the implementation of new ideas is in a way a chance occurrence; dependent upon the prevailing socioeconomic conditions, power structure and the currently existing paradigms in a given field of study at a given point in time. Introduction of new ideas or reintroduction of old, have been successfully accomplished, and can still be done!

A paradigm can be described as a theoretical framework for viewing a problem or situation. Our attempts to solve the problem are then directed at the principals of the paradigm. Thomas Kuhn in his book, *The Structure of Scientific Revolutions*, describes paradigms as follows: "These I take to be universally recognized scientific achievements **that for a time** provide model problems and solutions to a community of practitioners". He adds: "In a science, a paradigm is rarely an object for replication. Instead, like an accepted judicial decision in the common law, it is an object for further articulation and specification under new or more stringent conditions." And then: "**Paradigms change, but only after a crisis is reached**."

Healthcare has reached this crisis point. Today in America, healthcare is consuming over 15.3% or 1.67 trillion dollars of our nation's gross domestic product (GDP), and is growing rapidly. The current level is 4.3 times our nation's spending on national defense. Health insurance will cost a family of four an average of 14,500 dollars per year in 2006. Healthcare expenditures are taking over our lives and are robbing our children's future. The term healthcare is no longer descriptive of our healthcare profession.

The practice of the medical profession today might more appropriately be referred to as "illness-care" or "sickcare." Today's focus is not on preventing illness, but rather on diagnosing illness and then symptomatically offering treatments that are not directed at correcting the fundamental underlying cause of the illness. Where is prevention? I say, preventative medicine should be about optimizing your lifestyle for health and illness prevention — all of the time. Not going to doctors and obtaining diagnostic tests, waiting to discover if something is already wrong. Then, if nothing tests positive, keeping on with the same lifestyle and retesting annually until a problem does develop.

The word health has become deceptive. Health and wellness are considered ground zero today — a presumed absence of disease or a lack of symptoms — a neutral state. Today's "sickcare" system is focused on two states of being, life and death.

FIGURE 12.1 Conventional "Sickcare"

Is that good enough — should we settle for so little? Health and wellness are really positive states; ones that are better than average or superior states of being that are attained by an individual through proper attention to one's lifestyle. I believe "healthcare" should be a continuum that focuses on prevention and maintenance of health and wellness, rather than today's conventional "sickcare" paradigm which focuses on the health-illness-disability-death continuum.

As we quoted at the outset from Thomas Paine (*Common Sense*, 1776), "… a long habit of not thinking a thing wrong, gives it a superficial appearance of being right, and raises at first a formidable outcry in deference of custom. But the tumult soon subsides. Time makes more converts than reason."

The time has come for a paradigm shift, a change in direction — a Healthcare Revolution — focusing on the prevention of disease and the maintenance of a state of superior health! I believe "healthcare" needs to be redefined as being an optimal wellness-wellness-health-illness-disability-death continuum, with a preventative emphasis on health, wellness and optimal wellness.

FIGURE 12.2 Conventional Care — ConnexinCare

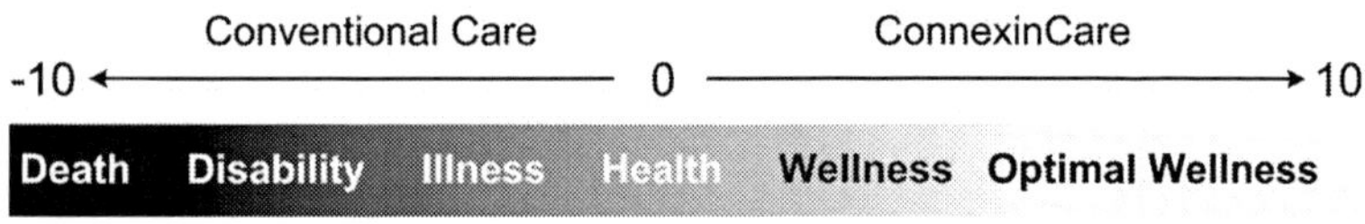

I advocate that conventional or traditional medicine place more emphasis on what I define as Molecular Medicine. Molecular Medicine is understanding health and disease at the molecular level and how this influences our entire body and mind. Then, using this information to promote health, prevent disease and achieve ultimate wellness. Empowering individuals to self-monitor current and future health indices, to enable objective tracking of one's state of being, in order to best influence their molecular makeup through diet, exercise, stress reduction and environment. Then, incorporating these results into a feed-back directed lifestyle aimed at achieving an optimal state of health and wellness.

Molecular Medicine is self-optimization of your natural genetic makeup, not externally directed manipulation or alteration. It is the medicine of the people, and requires self motivation, confidence, empowerment, self discipline and commitment.

Throughout the history of man many significant paradigm changes have taken place, most for the better — it can be done! Let us look at a few of the major paradigm shifts that have occurred in science throughout history. For thousands of years man believed the Earth to be flat, then in 300 BC a scientist named Eratosthenes in Alexandria, Egypt, did some experiments that proved the Earth to be round and calculated the circumference of the Earth within only a few percent of error. In 200 BC, Aristarchus of Samos proposed that the Sun was the center of the solar system. Later, Aristotle and Ptolemy put forth the idea of an earth-centered solar system that lasted for nearly 2,000 years. Then in the 16th century, Copernicus, fearful of peer and Church ridicule all of his life, waited until he was on his death bed to publish his work reestablishing the Sun as the center of the solar system. More recently, the concept that the universe started at time zero with the "big bang" was "proved" by Stephen Hawking and Roger Penrose in the 1960's. Some 2,000 years prior Aristotle proposed that time always existed and invoked the principal that out of nothing, nothing comes. Now in 2000, "string theory" refutes the "big bang" and we are back to an Aristotelian universe.

Healthcare is in desperate need of change. Clearly, we need to be focusing on prevention. As with all change, this can evoke fear and pain for some. Nonetheless, history has taught us that the only constant is change. In this book, *The Connexin Connection*, I present a new standard for healthcare, the Connexin Molecular Fitness Lifestyle; which will make us whole again by reinforcing the keystone to our health and life, the connexin molecule. It is interesting that the forefather of medicine, Hippocrates, in 400 B.C. believed in the unity of life and the singularity of disease. The time has come to revive his ancient proposal by embracing and acting upon our scientific knowledge of the connexin molecule.

I believe, restoration of the connexin molecule will herald a renaissance in our health, in our healthcare, and in the future of mankind.

Glossary

Acidosis

A state in the body where there is an actual or relative increase in acid in proportion to alkali or a decrease in alkali. Cellular function is impaired when acidosis exists.

Angstrom

A unit of length equal to one ten-billionth of a meter.

Animal Fat

Animal fats are fats or lipids that are of animal origin.

Apoptosis

Cell death by fragmentation into particles which are taken up by other cells.

Arachidonic Acid

Arachidonic acid is unsaturated fatty acid that is a precursor to inflammatory molecules such as prostaglandins.

Bone Density

Bone density is the amount of bone tissue in a certain volume of bone. This measurement is generally regarded as a measure of bone soundness, and if low a person is said to have osteoporosis.

Cellular

Relating to, composed of, or derived from cells.

Cholesterol Synthesis Pathway

The chemical pathway and steps taken to synthesize cholesterol in the body.

Connexin Dysfunction

Connexin dysfunction is abnormal function of connexin molecules.

Diet

Diet as defined in the context of this book is the foods and nutrients one consumes.

Down-Regulated

To down-regulate is to reduce the amount of protein produced or reduce its function. For example, connexin down-regulation is the reduction of connexin molecule production, membrane insertion or functionality.

Eight Diseases of Modern Man

The Eight Major Diseases of Modern Man are: cancer, heart disease, diabetes, arthritis, osteoporosis, Alzheimer's, hypothyroidism (obesity) and eye disease (cataracts).

Feed-Forward

A feed-forward system is such that one step enhances another in a positive direction.

Fitness

The condition of being suitable or qualified.

Fitness-related Health Performance

Fitness-related health performance is performance that is related to an organism's physical and mental fitness.

Functional Connexin

A connexin molecule that is able to perform its function of cellular communication both molecularly and electrically.

Gap Junction

A word synonomous with connexin which refers to the same protein.

Gating Properties

The properties that regulate the connexin pore size and the molecules that can pass through it.

Gene Expression

The process by which a gene's coded information is converted into the molecules present in a cell. For example, a connexin gene is expressed when its code is transcribed into mRNA and then translated into a protein.

Glycemic Index

The glycemic index is a measure of how much your blood sugar increases after eating a carbohydrate food.

Glycemic Load

The glycemic load is equal to the glycemic index of a food times the number of grams of carbohydrate in the food being eaten.

Healing Rate

The rate or speed at which a tissue heals.

Health

Health is a state of being that entails full connectivity of cells and resultant absence of disease and illness.

Health Transformation

The current state of our nation and world in which we are becoming a global society immersed in illness.

Hippocrates
Greek physician. Considered to be the father of medicine. 460-377 B.C.

Hypothyroidism
Diminished production of thyroid hormone or diminished responsiveness to thyroid hormone. May result in weight gain, fatigue, depression, decreased libido, and high cholesterol.

mRNA
This stands for messenger RNA (ribonucleic acid) the gene transcript that codes for the protein sequence of amino acids that make up a protein molecule.

Mental Performance
The ability to perform mental tasks over time.

Metabolic Acidosis
A state of decreased pH and bicarbonate content in bodily fluids that can be caused by excess consumption of acid generating foods and insufficient intake of bicarbonate or alkalinizing foods.

Molecule
Two or more atoms (carbon, oxygen, hydrogen, etc.) joined by a chemical bond.

Molecular Fitness

Enhanced function of the connexin molecule or molecular systems involved in connexin function.

Muscle Mass

Physical amount of muscle that is present.

Performance

The efficiency of a system while performing tasks. The operational and support characteristics of a system that allow it to effectively and efficiently perform a given mission over time.

Physical Performance

The ability to perform physical tasks over time.

Plato

Greek philosopher, 428-348 B.C.

PRAL

Potential renal acid load.

Rabbit Starvation

This is a condition that the early American settlers encountered when they ate an inadequate diet that was unbalanced and contained an excess of protein leading to illness and death.

Species

A species is a group of related organisms that share a distinctive form and are capable of interbreeding.

Up-regulated

Up-regulation is an actual or functional increase of a protein molecule. For example, connexin up-regulation is an increase of connexin molecule synthesis or function.

Wellness

Wellness is a dimension of health that is beyond the absence of disease, which includes superlative social, emotional, and spiritual aspects of health.

References

The following references, listed by chapter, were used in the creation of this book. If you would like to know more about the details of any or all of these scientific papers, visit www.HealthConnexin.com/library.

Chapter 1 — Introduction

1. Alonso A, de la FC, Martin-Arnau AM, De IJ, Martinez JA, & Martinez-Gonzalez MA (2004). Fruit and vegetable consumption is inversely associated with blood pressure in a Mediterranean population with a high vegetable-fat intake: the Seguimiento Universidad de Navarra (SUN) Study. *Br J Nutr* **92**, 311-319.

2. Currat M & Excoffier L (2004). Modern humans did not admix with Neanderthals during their range expansion into Europe. *PLoS Biol* **2**, e421.

3. Esposito K & Giugliano D (2002). Mediterranean diet and prevention of coronary heart disease. *J Endocrinol Invest* **25**, 296-299.

4. Esposito K, Marfella R, Ciotola M, Di PC, Giugliano F, Giugliano G, D'Armiento M, D'Andrea F, & Giugliano D (2004). Effect of a mediterranean-style diet on endothelial dysfunction and markers of vascular inflammation in the metabolic syndrome: a randomized trial. *JAMA* **292**, 1440-1446.

5. Halliday JL, Collins VR, Aitken MA, Richards MP, & Olsson CA (2004). Genetics and public health--evolution, or revolution? *J Epidemiol Community Health* **58**, 894-899.

6. Knoops KT, de Groot LC, Kromhout D, Perrin AE, Moreiras-Varela O, Menotti A, & van Staveren WA (2004). Mediterranean diet, lifestyle factors, and 10-year mortality in elderly European men and women: the HALE project. *JAMA* **292**, 1433-1439.

7. La VC (2004). Mediterranean diet and cancer. *Public Health Nutr* 7, 965-968.

8. Lalueza C, Perez-Perez A, & Turbon D (1996). Dietary inferences through buccal microwear analysis of middle and upper Pleistocene human fossils. *Am J Phys Anthropol* **100**, 367-387.

9. Lasheras C, Fernandez S, & Patterson AM (2000). Mediterranean diet and age with respect to overall survival in institutionalized, nonsmoking elderly people. *Am J Clin Nutr* **71**, 987-992.

10. Lieberman DE, McBratney BM, & Krovitz G (2002). The evolution and development of cranial form in Homosapiens. *Proc Natl Acad Sci U S A* **99**, 1134-1139.

11. Panza F, Solfrizzi V, Colacicco AM, D'Introno A, Capurso C, Torres F, Del PA, Capurso S, & Capurso A (2004). Mediterranean diet and cognitive decline. *Public Health Nutr* 7, 959-963.

12. Passarino G, Underhill PA, Cavalli-Sforza LL, Semino O, Pes GM, Carru C, Ferrucci L, Bonafe M, Franceschi C, Deiana L, Baggio G, & De BG (2001). Y chromosome binary markers to study the high prevalence of males in Sardinian centenarians and the genetic structure of the Sardinian population. *Hum Hered* **52**, 136-139.

13. Richards MP, Pettitt PB, Trinkaus E, Smith FH, Paunovic M, & Karavanic I (2000). Neanderthal diet at Vindija and Neanderthal predation: the evidence from stable isotopes. *Proc Natl Acad Sci U S A* **97**, 7663-7666.

14. Richards MP, Jacobi R, Cook J, Pettitt PB, & Stringer CB (2005). Isotope evidence for the intensive use of marine foods by Late Upper Palaeolithic humans. *J Hum Evol.*

15. Semino O, Passarino G, Oefner PJ, Lin AA, Arbuzova S, Beckman LE, De BG, Francalacci P, Kouvatsi A, Limborska S, Marcikiae M, Mika A, Mika B, Primorac D, Santachiara-Benerecetti AS, Cavalli-Sforza LL, & Underhill PA (2000). The genetic legacy of Paleolithic Homo sapiens sapiens in extant Europeans: a Y chromosome perspective. *Science* **290**, 1155-1159.

16. Trichopoulou A (2004). Traditional Mediterranean diet and longevity in the elderly: a review. *Public Health Nutr* 7, 943-947.

Chapter 2 — Connexins

1. Baker TS, Sosinsky GE, Caspar DL, Gall C, & Goodenough DA (1985). Gap junction structures. VII. Analysis of connexon images obtained with cationic and anionic negative stains. *J Mol Biol* **184**, 81-98.

2. Bennett MV (1994). Connexins in disease. *Nature* **368**, 18-19.

3. Bennett MV, Zheng X, & Sogin ML (1994). The connexins and their family tree. *Soc Gen Physiol Ser* **49**, 223-233.

4. De MA, Vega VL, & Contreras JE (2002). Gap junctions, homeostasis, and injury. *J Cell Physiol* **191**, 269-282.

5. Evans WH & Martin PE (2002). Gap junctions: structure and function (Review). *Mol Membr Biol* **19**, 121-136.

6. Fleishman SJ, Unger VM, Yeager M, & Ben-Tal N (2004). A Calpha model for the transmembrane alpha helices of gap junction intercellular channels. *Mol Cell* **15**, 879-888.

7. Goodenough DA & Revel JP (1970). A fine structural analysis of intercellular junctions in the mouse liver. *J Cell Biol* **45**, 272-290.

8. Goodenough DA (1974). Bulk isolation of mouse hepatocyte gap junctions. Characterization of the principal protein, connexin. *J Cell Biol* **61**, 557-563.

9. Kilarski WM, Severs, NJ, Gourdie, RG, Rezapour, M, Backstrom T, Roomans GM, Ulmsten U (1993). Gap junction density in human myometrium at term revealed by an anti-peptide antibody and laser scanning confocal microscopy . *Folia Histochem Cytobiol* **31**, 155-160.

10. Kilarski WM, Rezapour M, Backstrom T, Roomans GM, Ulmsten U (1994). Morphometric analysis of gap junction density in human myometrium at term. *Acta Obstet Gynecol Scand* 73, 377-384.

11. Kumar NM & Gilula NB (1992). Molecular biology and genetics of gap junction channels. *Semin Cell Biol* 3, 3-16.

12. Lopez P, Balicki D, Buehler LK, Falk MM, & Chen SC (2001). Distribution and dynamics of gap junction channels revealed in living cells. *Cell Commun Adhes* **8**, 237-242.

13. Sohl G & Willecke K (2004). Gap junctions and the connexin protein family. *Cardiovasc Res* **62**, 228-232.

14. Wei CJ, Xu X, & Lo CW (2004). Connexins and cell signaling in development and disease. *Annu Rev Cell Dev Biol* **20**, 811-838.

Chapter 3 — Connexins and disease

Statistics

1. Heart Disease and Stroke Statistics - 2005 Update. pp 3-31, 2005.
   Ref Type: Report

2. American Diabetes Association Statistics 2002. 2005.
   Ref Type: Report

3. American Cancer Society. Cancer Facts and Figures 2005. 1-39. 2005.
   Ref Type: Report

4. Alzheimer's Association. Statistics about Alzheimer's Disease. 2005.
   Ref Type: Report

5. Canaris GJ, Manowitz NR, Mayor G, & Ridgway EC (2000). The Colorado thyroid disease prevalence study. *Arch Intern Med* **160**, 526-534.

6. Congdon N, O'Colmain B, Klaver CC, Klein R, Munoz B, Friedman DS, Kempen J, Taylor HR, & Mitchell P (2004). Causes and prevalence of visual impairment among adults in the United States. *Arch Ophthalmol* **122**, 477-485.

7. Friedman DS, O'Colmain BJ, Munoz B, Tomany SC, McCarty C, de Jong PT, Nemesure B, Mitchell P, & Kempen J (2004). Prevalence of age-related macular degeneration in the United States. *Arch Ophthalmol* **122**, 564-572.

8. National Center for Disease Control - National Center for Chronic Disease Prevention and Health Promotion. Arthritis. 2005.
   Ref Type: Report

9. National Osteoporosis Foundation. America's Bone Health: The State of Osteoporosis. 2005.
Ref Type: Report

HEART

1. de G, Jr. & Coronel R (2004). Acute ischemia-induced gap junctional uncoupling and arrhythmogenesis. *Cardiovasc Res* **62**, 323-334.

2. de WC, Roos F, Bolz SS, & Pohl U (2003). Lack of vascular connexin 40 is associated with hypertension and irregular arteriolar vasomotion. *Physiol Genomics* **13**, 169-177.

3. Gutstein DE, Morley GE, & Fishman GI (2001). Conditional gene targeting of connexin43: exploring the consequences of gap junction remodeling in the heart. *Cell Commun Adhes* **8**, 345-348.

4. Gutstein DE, Morley GE, Tamaddon H, Vaidya D, Schneider MD, Chen J, Chien KR, Stuhlmann H, & Fishman GI (2001). Conduction slowing and sudden arrhythmic death in mice with cardiac-restricted inactivation of connexin43. *Circ Res* **88**, 333-339.

5. Morley GE, Danik SB, Bernstein S, Sun Y, Rosner G, Gutstein DE, & Fishman GI (2005). Reduced intercellular coupling leads to paradoxical propagation across the Purkinje-ventricular junction and aberrant myocardial activation. *Proc Natl Acad Sci U S A* **102**, 4126-4129.

6. Poelzing S & Rosenbaum DS (2004). Altered connexin43 expression produces arrhythmia substrate in heart failure. *Am J Physiol Heart Circ Physiol* **287**, H1762-H1770.

7. Severs NJ (2001). Gap junction remodeling and cardiac arrhythmogenesis: cause or coincidence? *J Cell Mol Med* **5**, 355-366.

8. Severs NJ, Coppen SR, Dupont E, Yeh HI, Ko YS, & Matsushita T (2004). Gap junction alterations in human cardiac disease. *Cardiovasc Res* **62**, 368-377.

9. Simon AM, McWhorter AR, Dones JA, Jackson CL, & Chen H (2004). Heart and head defects in mice lacking pairs of connexins. *Dev Biol* **265**, 369-383.

10. Yeh HI, Lu CS, Wu YJ, Chen CC, Hong RC, Ko YS, Shiao MS, Severs NJ, & Tsai CH (2003). Reduced expression of endothelial connexin37 and connexin40 in hyperlipidemic mice: recovery of connexin37 after 7-day simvastatin treatment. *Arterioscler Thromb Vasc Biol* **23**, 1391-1397.

Cancer

1. Bertram JS (1993). Cancer prevention by carotenoids. Mechanistic studies in cultured cells. *Ann N Y Acad Sci* **691**, 177-191.

2. Bertram JS (2004). Dietary carotenoids, connexins and cancer: what is the connection? *Biochem Soc Trans* **32**, 985-989.

3. Bertram JS (2004). Induction of connexin 43 by carotenoids: functional consequences. *Arch Biochem Biophys* **430**, 120-126.

4. Cesen-Cummings K, Fernstrom MJ, Malkinson AM, & Ruch RJ (1998). Frequent reduction of gap junctional intercellular communication and connexin43 expression in human and mouse lung carcinoma cells. *Carcinogenesis* **19**, 61-67.

5. Chen Y, Huhn D, Knosel T, Pacyna-Gengelbach M, Deutschmann N, & Petersen I (2005). Downregulation of connexin 26 in human lung cancer is related to promoter methylation. *Int J Cancer* **113**, 14-21.

6. Cochand-Priollet B, Raison D, Molinie V, Guillausseau PJ, Wassef M, & Bouchaud C (1998). Altered gap and tight junctions in human thyroid oncocytic tumors: a study of 8 cases by freeze-fracture. *Ultrastruct Pathol* **22**, 413-420.

7. Fernstrom MJ, Koffler LD, bou-Rjaily G, Boucher PD, Shewach DS, & Ruch RJ (2002). Neoplastic reversal of human ovarian carcinoma cells transfected with connexin43. *Exp Mol Pathol* 73, 54-60.

8. Gee J, Tanaka M, & Grossman HB (2003). Connexin 26 is abnormally expressed in bladder cancer. *J Urol* **169**, 1135-1137.

9. Goldberg GS & Bertram JS (1994). Retinoids, gap junctional communication and suppression of epithelial tumors. *In Vivo* **8**, 745-754.

10. Grossman HB, Liebert M, Lee IW, & Lee SW (1994). Decreased connexin expression and intercellular communication in human bladder cancer cells. *Cancer Res* **54**, 3062-3065.

11. Hanna EA, Umhauer S, Roshong SL, Piechocki MP, Fernstrom MJ, Fanning JD, & Ruch RJ (1999). Gap junctional intercellular communication and connexin43 expression in human ovarian surface epithelial cells and ovarian carcinomas in vivo and in vitro. *Carcinogenesis* **20**, 1369-1373.

12. Huang RP, Hossain MZ, Sehgal A, & Boynton AL (1999). Reduced connexin43 expression in high-grade human brain glioma cells. *J Surg Oncol* 70, 21-24.

13. King TJ, Fukushima LH, Hieber AD, Shimabukuro KA, Sakr WA, & Bertram JS (2000). Reduced levels of connexin43 in cervical dysplasia: inducible expression in a cervical carcinoma cell line decreases neoplastic potential with implications for tumor progression. *Carcinogenesis* **21**, 1097-1109.

14. Mesnil M (2002). Connexins and cancer. *Biol Cell* **94**, 493-500.

15. Pu P, Xia Z, Yu S, & Huang Q (2004). Altered expression of Cx43 in astrocytic tumors. *Clin Neurol Neurosurg* **107**, 49-54.

16. Ruch RJ, Cesen-Cummings K, & Malkinson AM (1998). Role of gap junctions in lung neoplasia. *Exp Lung Res* **24**, 523-539.

17. Ruch RJ, Porter S, Koffler LD, Dwyer-Nield LD, & Malkinson AM (2001). Defective gap junctional intercellular communication in lung cancer: loss of an important mediator of tissue homeostasis and phenotypic regulation. *Exp Lung Res* **27**, 231-243.

18. Sheen IS, Jeng KS, Wang PC, Shih SC, Chang WH, Wang HY, Chen CC, & Shyung LR (2004). Are gap junction gene connexins 26, 32 and 43 of prognostic values in hepatocellular carcinoma? A prospective study. *World J Gastroenterol* **10**, 2785-2790.

19. Trosko JE & Ruch RJ (1998). Cell-cell communication in carcinogenesis. *Front Biosci* 3, d208-d236.

20. Vine AL & Bertram JS (2002). Cancer chemoprevention by connexins. *Cancer Metastasis Rev* 21, 199-216.

21. Vine AL, Leung YM, & Bertram JS (2005). Transcriptional regulation of connexin 43 expression by retinoids and carotenoids: similarities and differences. *Mol Carcinog* **43**, 75-85.

22. Zhang LX, Cooney RV, & Bertram JS (1992). Carotenoids upregulate connexin43 gene expression independent of their provitamin A or antioxidant properties. *Cancer Res* **52**, 5707-5712.

Diabetes

1. Caton D, Calabrese A, Mas C, Serre-Beinier V, Wonkam A, & Meda P (2002). Beta-cell crosstalk: a further dimension in the stimulus-secretion coupling of glucose-induced insulin release. *Diabetes Metab* **28**, 3S45-3S53.

2. Charollais A, Gjinovci A, Huarte J, Bauquis J, Nadal A, Martin F, Andreu E, Sanchez-Andres JV, Calabrese A, Bosco D, Soria B, Wollheim CB, Herrera PL, & Meda P (2000). Junctional communication of pancreatic beta cells contributes to the control of insulin secretion and glucose tolerance. *J Clin Invest* **106**, 235-243.

3. Inoguchi T, Yu HY, Imamura M, Kakimoto M, Kuroki T, Maruyama T, & Nawata H (2001). Altered gap junction activity in cardiovascular tissues of diabetes. *Med Electron Microsc* **34**, 86-91.

4. Kistler J, Lin JS, Bond J, Green C, Eckert R, Merriman R, Tunstall M, & Donaldson P (1999). Connexins in the lens: are they to blame in diabetic cataractogenesis? *Novartis Found Symp* **219**, 97-108.

5. Kuroki T, Inoguchi T, Umeda F, Ueda F, & Nawata H (1998). High glucose induces alteration of gap junction permeability and phosphorylation of connexin-43 in cultured aortic smooth muscle cells. *Diabetes* **47**, 931-936.

6. Le GS, Martin D, Formenton A, Maechler P, Caille D, Waeber G, Meda P, & Haefliger JA (2003). Connexin-36 contributes to control function of insulin-producing cells. *J Biol Chem* **278**, 37690-37697.

7. Meda P (1996). [Role of intercellular communication via gap junctions in insulin secretion]. *Ann Endocrinol (Paris)* **57**, 481-483.

8. Meda P (1996). Gap junction involvement in secretion: the pancreas experience. *Clin Exp Pharmacol Physiol* **23**, 1053-1057.

9. Meda P (2003). Cx36 involvement in insulin secretion: characteristics and mechanism. *Cell Commun Adhes* **10**, 431-435.

10. Okruhlicova L, Tribulova N, Misejkova M, Kucka M, Stetka R, Slezak J, & Manoach M (2002). Gap junction remodelling is involved in the susceptibility of diabetic rats to hypokalemia-induced ventricular fibrillation. *Acta Histochem* **104**, 387-391.

11. Oku H, Kodama T, Sakagami K, & Puro DG (2001). Diabetes-induced disruption of gap junction pathways within the retinal microvasculature. *Invest Ophthalmol Vis Sci* **42**, 1915-1920.

12. Pitre DA, Seifert JL, & Bauer JA (2001). Perineurium inflammation and altered connexin isoform expression in a rat model of diabetes related peripheral neuropathy. *Neurosci Lett* **303**, 67-71.

13. Quesada I, Fuentes E, Andreu E, Meda P, Nadal A, & Soria B (2003). On-line analysis of gap junctions reveals more efficient electrical than dye coupling between islet cells. *Am J Physiol Endocrinol Metab* **284**, E980-E987.

14. Serre-Beinier V, Le GS, Belluardo N, Trovato-Salinaro A, Charollais A, Haefliger JA, Condorelli DF, & Meda P (2000). Cx36 preferentially connects beta-cells within pancreatic islets. *Diabetes* **49**, 727-734.

15. Vozzi C, Ullrich S, Charollais A, Philippe J, Orci L, & Meda P (1995). Adequate connexin-mediated coupling is required for proper insulin production. *J Cell Biol* **131**, 1561-1572.

ARTHRITIS

1. D'Andrea P & Vittur F (1996). Gap junctions mediate intercellular calcium signalling in cultured articular chondrocytes. *Cell Calcium* **20**, 389-397.

2. Kolomytkin OV, Marino AA, Sadasivan KK, Meek WD, Wolf RE, Hall V, McCarthy KJ, & Albright JA (2000). Gap junctions in human synovial cells and tissue. *J Cell Physiol* **184**, 110-117.

3. Schwab W, Hofer A, & Kasper M (1998). Immunohistochemical distribution of connexin 43 in the cartilage of rats and mice. *Histochem J* **30**, 413-419.

4. Tonon R & D'Andrea P (2000). Interleukin-1beta increases the functional expression of connexin 43 in articular chondrocytes: evidence for a Ca2+-dependent mechanism. *J Bone Miner Res* **15**, 1669-1677.

OSTEOPOROSIS

1. Donahue HJ, McLeod KJ, Rubin CT, Andersen J, Grine EA, Hertzberg EL, & Brink PR (1995). Cell-to-cell communication in osteoblastic networks: cell line-dependent hormonal regulation of gap junction function. *J Bone Miner Res* **10**, 881-889.

2. Furlan F, Lecanda F, Screen J, & Civitelli R (2001). Proliferation, differentiation and apoptosis in connexin43-null osteoblasts. *Cell Commun Adhes* **8**, 367-371.

3. Ilvesaro J & Tuukkanen J (2003). Gap-junctional regulation of osteoclast function. *Crit Rev Eukaryot Gene Expr* **13**, 133-146.

4. Lecanda F, Towler DA, Ziambaras K, Cheng SL, Koval M, Steinberg TH, & Civitelli R (1998). Gap junctional communication modulates gene expression in osteoblastic cells. *Mol Biol Cell* **9**, 2249-2258.

5. Lecanda F, Warlow PM, Sheikh S, Furlan F, Steinberg TH, & Civitelli R (2000). Connexin43 deficiency causes delayed ossification, craniofacial abnormalities, and osteoblast dysfunction. *J Cell Biol* **151**, 931-944.

6. Li Z, Zhou Z, Yellowley CE, & Donahue HJ (1999). Inhibiting gap junctional intercellular communication alters expression of differentiation markers in osteoblastic cells. *Bone* **25**, 661-666.

7. Mitchell JA, Ou C, Chen Z, Nishimura T, & Lye SJ (2001). Parathyroid hormone-induced up-regulation of connexin-43 messenger ribonucleic acid (mRNA) is mediated by sequences within both the promoter and the 3'untranslated region of the mRNA. *Endocrinology* **142**, 907-915.

8. Nakaoka R, Ahmed S, & Tsuchiya T (2005). Hydroxy apatite microspheres enhance gap junctional intercellular communication of human osteoblasts composed of connexin 43 and 45. *J Biomed Mater Res A* **74**, 181-186.

9. Plotkin LI, Manolagas SC, & Bellido T (2002). Transduction of cell survival signals by connexin-43 hemichannels. *J Biol Chem* **277**, 8648-8657.

10. Ransjo M, Sahli J, & Lie A (2003). Expression of connexin 43 mRNA in microisolated murine osteoclasts and regulation of bone resorption in vitro by gap junction inhibitors. *Biochem Biophys Res Commun* **303**, 1179-1185.

11. Schiller PC, Mehta PP, Roos BA, & Howard GA (1992). Hormonal regulation of intercellular communication: parathyroid hormone increases connexin 43 gene expression and gap-junctional communication in osteoblastic cells. *Mol Endocrinol* **6**, 1433-1440.

12. Schiller PC, Roos BA, & Howard GA (1997). Parathyroid hormone up-regulation of connexin 43 gene expression in osteoblasts depends on cell phenotype. *J Bone Miner Res* **12**, 2005-2013.

13. Schiller PC, D'Ippolito G, Balkan W, Roos BA, & Howard GA (2001). Gap-junctional communication mediates parathyroid hormone stimulation of mineralization in osteoblastic cultures. *Bone* **28**, 38-44.

14. Stains JP & Civitelli R (2005). Cell-cell interactions in regulating osteogenesis and osteoblast function. *Birth Defects Res C Embryo Today* **75**, 72-80.

15. Stains JP & Civitelli R (2005). Gap junctions regulate extracellular signal-regulated kinase signaling to affect gene transcription. *Mol Biol Cell* **16**, 64-72.

16. Yamaguchi DT & Ma D (2003). Mechanism of pH regulation of connexin 43 expression in MC3T3-E1 cells. *Biochem Biophys Res Commun* **304**, 736-739.

17. Zhang D, Weinbaum S, & Cowin SC (1998). Electrical signal transmission in a bone cell network: the influence of a discrete gap junction. *Ann Biomed Eng* **26**, 644-659.

18. Ziambaras K, Lecanda F, Steinberg TH, & Civitelli R (1998). Cyclic stretch enhances gap junctional communication between osteoblastic cells. *J Bone Miner Res* **13**, 218-228

Alzheimer's

1. Butkevich E, Hulsmann S, Wenzel D, Shirao T, Duden R, & Majoul I (2004). Drebrin is a novel connexin-43 binding partner that links gap junctions to the submembrane cytoskeleton. *Curr Biol* **14**, 650-658

Thyroid

1. Flachon V, Durand C, Selmi-Ruby S, Rousset B, & Munari-Silem Y (2001). Connexin 32 fused to the green fluorescent protein retains its ability to control the proliferation of thyroid cells. *Cell Commun Adhes* **8**, 447-452.

2. Flachon V, Tonoli H, Selmi-Ruby S, Durand C, Rabilloud R, Rousset B, & Munari-Silem Y (2002). Thyroid cell proliferation in response to forced expression of gap junction proteins. *Eur J Cell Biol* **81**, 243-252.

3. Green LM, Lazarus JP, LaBue M, & Shah MM (1995). Reduced cell-cell communication in a spontaneous murine model of autoimmune thyroid disease. *Endocrinology* **136**, 3611-3618.

4. Green LM, LaBue M, Lazarus JP, & Jennings JC (1996). Reduced cell-cell communication in experimentally induced autoimmune thyroid disease. *Endocrinology* **137**, 2823-2832.

5. Kolaja KL, Petrick JS, & Klaassen CD (2000). Inhibition of gap-junctional-intercellular communication in thyroid-follicular cells by propylthiouracil and low iodine diet. *Toxicology* **143**, 195-202.

6. Munari-Silem Y, Audebet C, & Rousset B (1991). Hormonal control of cell to cell communication: regulation by thyrotropin of the gap junction-mediated dye transfer between thyroid cells. *Endocrinology* **128**, 3299-3309.

7. Munari-Silem Y, Guerrier A, Fromaget C, Rabilloud R, Gros D, & Rousset B (1994). Differential control of connexin-32 and connexin-43 expression in thyroid epithelial cells: evidence for a direct relationship between connexin-32 expression and histiotypic morphogenesis. *Endocrinology* **135**, 724-734.

8. Munari-Silem Y (1996). [Gap junctions in the thyroid gland: distribution, regulation, function]. *Ann Endocrinol (Paris)* **57**, 484-486.

9. Statuto M, Audebet C, Tonoli H, Selmi-Ruby S, Rousset B, & Munari-Silem Y (1997). Restoration of cell-to-cell communication in thyroid cell lines by transfection with and stable expression of the connexin-32 gene. Impact on cell proliferation and tissue-specific gene expression. *J Biol Chem* **272**, 24710-24716.

10. Stock A, Sies H, & Stahl W (1998). Enhancement of gap junctional communication and connexin43 expression by thyroid hormones. *Biochem Pharmacol* **55**, 475-479.

11. Stock A & Sies H (2000). Thyroid hormone receptors bind to an element in the connexin43 promoter. *Biol Chem* **381**, 973-979.

Ocular

1. Baruch A, Greenbaum D, Levy ET, Nielsen PA, Gilula NB, Kumar NM, & Bogyo M (2001). Defining a link between gap junction communication, proteolysis, and cataract formation. *J Biol Chem* **276**, 28999-29006.

2. Berry V, Mackay D, Khaliq S, Francis PJ, Hameed A, Anwar K, Mehdi SQ, Newbold RJ, Ionides A, Shiels A, Moore T, & Bhattacharya SS (1999). Connexin 50 mutation in a family with congenital "zonular nuclear" pulverulent cataract of Pakistani origin. *Hum Genet* **105**, 168-170.

3. Berthoud VM, Montegna EA, Atal N, Aithal NH, Brink PR, & Beyer EC (2001). Heteromeric connexons formed by the lens connexins, connexin43 and connexin56. *Eur J Cell Biol* **80**, 11-19.

4. Brink PR, Valiunas V, Moore L, Birzgalis A, & Walcott B (2002). The role of gap junctions in lacrimal acinar cells: the formation of tears. *Adv Exp Med Biol* **506**, 109-113.

5. Eckert R (2002). pH gating of lens fibre connexins. *Pflugers Arch* **443**, 843-851.

6. Fernandes R, Girao H, & Pereira P (2004). High glucose down-regulates intercellular communication in retinal endothelial cells by enhancing degradation of connexin 43 by a proteasome-dependent mechanism. *J Biol Chem* **279**, 27219-27224.

7. Gao J, Sun X, Martinez-Wittinghan FJ, Gong X, White TW, & Mathias RT (2004). Connections between connexins, calcium, and cataracts in the lens. *J Gen Physiol* **124**, 289-300.

8. Goodenough DA (1992). The crystalline lens. A system networked by gap junctional intercellular communication. *Semin Cell Biol* **3**, 49-58.

9. Li AF, Sato T, Haimovici R, Okamoto T, & Roy S (2003). High glucose alters connexin 43 expression and gap junction intercellular communication activity in retinal pericytes. *Invest Ophthalmol Vis Sci* **44**, 5376-5382.

10. Lin D, Boyle DL, & Takemoto DJ (2003). IGF-I-induced phosphorylation of connexin 43 by PKCgamma: regulation of gap junctions in rabbit lens epithelial cells. *Invest Ophthalmol Vis Sci* **44**, 1160-1168.

11. Mackay D, Ionides A, Kibar Z, Rouleau G, Berry V, Moore A, Shiels A, & Bhattacharya S (1999). Connexin46 mutations in autosomal dominant congenital cataract. *Am J Hum Genet* **64**, 1357-1364.

12. Martinez-Wittinghan FJ, Sellitto C, Li L, Gong X, Brink PR, Mathias RT, & White TW (2003). Dominant cataracts result from incongruous mixing of wild-type lens connexins. *J Cell Biol* **161**, 969-978.

13. Martinez-Wittinghan FJ, Sellitto C, White TW, Mathias RT, Paul D, & Goodenough DA (2004). Lens gap junctional coupling is modulated by connexin identity and the locus of gene expression. *Invest Ophthalmol Vis Sci* **45**, 3629-3637.

14. McMahon DG, Knapp AG, & Dowling JE (1989). Horizontal cell gap junctions: single-channel conductance and modulation by dopamine. *Proc Natl Acad Sci U S A* **86**, 7639-7643.

15. Montgomery JR, White TW, Martin BL, Turner ML, & Holland SM (2004). A novel connexin 26 gene mutation associated with features of the keratitis-ichthyosis-deafness syndrome and the follicular occlusion triad. *J Am Acad Dermatol* **51**, 377-382.

16. Oku H, Kodama T, Sakagami K, & Puro DG (2001). Diabetes-induced disruption of gap junction pathways within the retinal microvasculature. *Invest Ophthalmol Vis Sci* **42**, 1915-1920.

17. Pal JD, Berthoud VM, Beyer EC, Mackay D, Shiels A, & Ebihara L (1999). Molecular mechanism underlying a Cx50-linked congenital cataract. *Am J Physiol* **276**, C1443-C1446.

18. Pal JD, Liu X, Mackay D, Shiels A, Berthoud VM, Beyer EC, & Ebihara L (2000). Connexin46 mutations linked to congenital cataract show loss of gap junction channel function. *Am J Physiol Cell Physiol* **279**, C596-C602.

19. Rong P, Wang X, Niesman I, Wu Y, Benedetti LE, Dunia I, Levy E, & Gong X (2002). Disruption of Gja8 (alpha8 connexin) in mice leads to microphthalmia associated with retardation of lens growth and lens fiber maturation. *Development* **129**, 167-174.

20. Sellitto C, Li L, & White TW (2004). Connexin50 is essential for normal postnatal lens cell proliferation. *Invest Ophthalmol Vis Sci* **45**, 3196-3202.

21. Shiels A, Mackay D, Ionides A, Berry V, Moore A, & Bhattacharya S (1998). A missense mutation in the human connexin50 gene (GJA8) underlies autosomal dominant "zonular pulverulent" cataract, on chromosome 1q. *Am J Hum Genet* **62**, 526-532.

22. Walcott B, Moore LC, Birzgalis A, Claros N, Valiunas V, Ott T, Willecke K, & Brink PR (2002). Role of gap junctions in fluid secretion of lacrimal glands. *Am J Physiol Cell Physiol* **282**, C501-C507.

Chapter 4 — Connexinx and the food we eat — I connexin closers / function reducers

1. Agrawal R & Daniel EE (1986). Control of gap junction formation in canine trachea by arachidonic acid metabolites. *Am J Physiol* **250**, C495-C505.

2. Boger DL, Sato H, Lerner AE, Guan X, & Gilula NB (1999). Arachidonic acid amide inhibitors of gap junction cell-cell communication. *Bioorg Med Chem Lett* **9**, 1151-1154.

3. Duffy HS, Ashton AW, O'Donnell P, Coombs W, Taffet SM, Delmar M, & Spray DC (2004). Regulation of connexin43 protein complexes by intracellular acidification. *Circ Res* **94**, 215-222.

4. Eckert R (2002). pH gating of lens fibre connexins. *Pflugers Arch* **443**, 843-851.

5. Ek-Vitorin JF, Calero G, Morley GE, Coombs W, Taffet SM, & Delmar M (1996). PH regulation of connexin43: molecular analysis of the gating particle. *Biophys J* **71**, 1273-1284.

6. Fernandes R, Girao H, & Pereira P (2004). High glucose down-regulates intercellular communication in retinal endothelial cells by enhancing degradation of connexin 43 by a proteasome-dependent mechanism. *J Biol Chem* **279**, 27219-27224.

7. Francis D, Stergiopoulos K, Ek-Vitorin JF, Cao FL, Taffet SM, & Delmar M (1999). Connexin diversity and gap junction regulation by pHi. *Dev Genet* **24**, 123-136.

8. Frassetto LA, Morris RC, Jr., & Sebastian A (1996). Effect of age on blood acid-base composition in adult humans: role of age-related renal functional decline. *Am J Physiol* **271**, F1114-F1122.

9. Frassetto LA, Todd KM, Morris RC, Jr., & Sebastian A (1998). Estimation of net endogenous noncarbonic acid production in humans from diet potassium and protein contents. *Am J Clin Nutr* **68**, 576-583.

10. Frassetto LA, Todd KM, Morris RC, Jr., & Sebastian A (2000). Worldwide incidence of hip fracture in elderly women: relation to consumption of animal and vegetable foods. *J Gerontol A Biol Sci Med Sci* **55**, M585-M592.

11. Giaume C, Randriamampita C, & Trautmann A (1989). Arachidonic acid closes gap junction channels in rat lacrimal glands. *Pflugers Arch* **413**, 273-279.

12. In't Veld PA, Pipeleers DG, & Gepts W (1986). Glucose alters configuration of gap junctions between pancreatic islet cells. *Am J Physiol* **251**, C191-C196.

13. Ince BA, Anderson EJ, & Neer RM (2004). Lowering dietary protein to U.S. Recommended dietary allowance levels reduces urinary calcium excretion and bone resorption in young women. *J Clin Endocrinol Metab* **89**, 3801-3807.

14. Inoguchi T, Ueda F, Umeda F, Yamashita T, & Nawata H (1995). Inhibition of intercellular communication via gap junction in cultured aortic endothelial cells by elevated glucose and phorbol ester. *Biochem Biophys Res Commun* **208**, 492-497.

15. Krishna G, Sleigh JW, & Rahman H (1998). Physiological predictors of death in exsanguinating trauma patients undergoing conventional trauma surgery. *Aust N Z J Surg* **68**, 826-829.

16. Kuroki T, Inoguchi T, Umeda F, Ueda F, & Nawata H (1998). High glucose induces alteration of gap junction permeability and phosphorylation of connexin-43 in cultured aortic smooth muscle cells. *Diabetes* **47**, 931-936.

17. Kurtz I, Maher T, Hulter HN, Schambelan M, & Sebastian A (1983). Effect of diet on plasma acid-base composition in normal humans. *Kidney Int* **24**, 670-680.

18. Lau AF (2005). c-Src: bridging the gap between phosphorylation- and acidification-induced gap junction channel closure. *Sci STKE* **2005**, e33.

19. Li AF, Sato T, Haimovici R, Okamoto T, & Roy S (2003). High glucose alters connexin 43 expression and gap junction intercellular communication activity in retinal pericytes. *Invest Ophthalmol Vis Sci* **44**, 5376-5382.

20. Lutz J (1984). Calcium balance and acid-base status of women as affected by increased protein intake and by sodium bicarbonate ingestion. *Am J Clin Nutr* **39**, 281-288.

21. Martinez AD & Saez JC (1999). Arachidonic acid-induced dye uncoupling in rat cortical astrocytes is mediated by arachidonic acid byproducts. *Brain Res* **816**, 411-423.

22. Miyachi E, Kato C, & Nakaki T (1994). Arachidonic acid blocks gap junctions between retinal horizontal cells. *Neuroreport* **5**, 485-488.

23. Morley GE, Taffet SM, & Delmar M (1996). Intramolecular interactions mediate pH regulation of connexin43 channels. *Biophys J* **70**, 1294-1302.

24. Morley GE, Ek-Vitorin JF, Taffet SM, & Delmar M (1997). Structure of connexin43 and its regulation by pHi. *J Cardiovasc Electrophysiol* **8**, 939-951.

25. Remer T & Manz F (1995). Potential renal acid load of foods and its influence on urine pH. *J Am Diet Assoc* **95**, 791-797.

26. Remer T (2000). Influence of diet on acid-base balance. *Semin Dial* **13**, 221-226.

27. Sato T, Haimovici R, Kao R, Li AF, & Roy S (2002). Downregulation of connexin 43 expression by high glucose reduces gap junction activity in microvascular endothelial cells. *Diabetes* **51**, 1565-1571.

28. Schmilinsky-Fluri G, Valiunas V, Willi M, & Weingart R (1997). Modulation of cardiac gap junctions: the mode of action of arachidonic acid. *J Mol Cell Cardiol* **29**, 1703-1713.

29. Sebastian A, Frassetto LA, Sellmeyer DE, Merriam RL, & Morris RC, Jr. (2002). Estimation of the net acid load of the diet of ancestral preagricultural Homo sapiens and their hominid ancestors. *Am J Clin Nutr* **76**, 1308-1316.

30. Sherman HC & Sinclair (1907). *J Biol Chem* **3**, 307.

31. Sorgen PL, Duffy HS, Spray DC, & Delmar M (2004). pH-dependent dimerization of the carboxyl terminal domain of Cx43. *Biophys J* **87**, 574-581.

32. Stalmans P & Himpens B (1997). Effect of increasing glucose concentrations and protein phosphorylation on intercellular communication in cultured rat retinal pigment epithelial cells. *Invest Ophthalmol Vis Sci* **38**, 1598-1609.

33. Trexler EB, Bukauskas FF, Bennett MV, Bargiello TA, & Verselis VK (1999). Rapid and direct effects of pH on connexins revealed by the connexin46 hemichannel preparation. *J Gen Physiol* **113**, 721-742.

34. Weingart R & Bukauskas FF (1998). Long-chain n-alkanols and arachidonic acid interfere with the Vm-sensitive gating mechanism of gap junction channels. *Pflugers Arch* **435**, 310-319.

35. Yamaguchi DT, Huang JT, & Ma D (1995). Regulation of gap junction intercellular communication by pH in MC3T3-E1 osteoblastic cells. *J Bone Miner Res* **10**, 1891-1899.

36. Yamaguchi DT & Ma D (2003). Mechanism of pH regulation of connexin 43 expression in MC3T3-E1 cells. *Biochem Biophys Res Commun* **304**, 736-739.

Chapter 5 — Connexins and the food we eat — II
connexin openers / function enhancers

1. Aust O, le-Agha N, Zhang L, Wollersen H, Sies H, & Stahl W (2003). Lycopene oxidation product enhances gap junctional communication. *Food Chem Toxicol* **41**, 1399-1407.

2. Bertram JS (1993). Inhibition of chemically induced neoplastic transformation by carotenoids. Mechanistic studies. *Ann N Y Acad Sci* **686**, 161-175.

3. Bertram JS & Bortkiewicz H (1995). Dietary carotenoids inhibit neoplastic transformation and modulate gene expression in mouse and human cells. *Am J Clin Nutr* **62**, 1327S-1336S.

4. Chappell RL, Zakevicius J, & Ripps H (2003). Zinc modulation of hemichannel currents in Xenopus oocytes. *Biol Bull* **205**, 209-211.

5. Chaumontet C, Bex V, Gaillard-Sanchez I, Seillan-Heberden C, Suschetet M, & Martel P (1994). Apigenin and tangeretin enhance gap junctional intercellular communication in rat liver epithelial cells. *Carcinogenesis* **15**, 2325-2330.

6. Chaumontet C, Droumaguet C, Bex V, Heberden C, Gaillard-Sanchez I, & Martel P (1997). Flavonoids (apigenin, tangeretin) counteract tumor promoter-induced inhibition of intercellular communication of rat liver epithelial cells. *Cancer Lett* **114**, 207-210.

7. Chen L, Wert SE, Hendrix EM, Russell PT, Cannon M, & Larsen WJ (1990). Hyaluronic acid synthesis and gap junction endocytosis are necessary for normal expansion of the cumulus mass. *Mol Reprod Dev* **26**, 236-247.

8. Chiba H, Sawada N, Oyamada M, Kojima T, Iba K, Ishii S, & Mori M (1994). Hormonal regulation of connexin 43 expression and gap junctional communication in human osteoblastic cells. *Cell Struct Funct* **19**, 173-177.

9. Clairmont A, Tessman D, Stock A, Nicolai S, Stahl W, & Sies H (1996). Induction of gap junctional intercellular communication by vitamin D in human skin fibroblasts is dependent on the nuclear Induction of gap junctional intercellular communication by vitamin D in human skin fibroblasts is dependent on the nuclear vitamin D receptor. *Carcinogenesis* **17**, 1389-1391.

10. Clairmont A & Sies H (1997). Evidence for a posttranscriptional effect of retinoic acid on connexin43 gene expression via the 3'-untranslated region. *FEBS Lett* **419**, 268-270.

11. Cordain L, Eaton SB, Sebastian A, Mann N, Lindeberg S, Watkins BA, O'Keefe JH, & Brand-Miller J (2005). Origins and evolution of the Western diet: health implications for the 21st century. *Am J Clin Nutr* **81**, 341-354.

12. Elias PM, Grayson S, Gross EG, Peck GL, & McNutt NS (1981). Influence of topical and systemic retinoids on basal cell carcinoma cell membranes. *Cancer* **48**, 932-938.

13. Fujita M, Spray DC, Choi H, Saez JC, Watanabe T, Rosenberg LC, Hertzberg EL, & Reid LM (1987). Glycosaminoglycans and proteoglycans induce gap junction expression and restore transcription of tissue-specific mRNAs in primary liver cultures. *Hepatology* 7, 1S-9S.

14. le-Agha N, Stahl W, & Sies H (2002). (-)-Epicatechin effects in rat liver epithelial cells: stimulation of gap junctional communication and counteraction of its loss due to the tumor promoter 12-O-tetradecanoylphorbol-13-acetate. *Biochem Pharmacol* **63**, 2145-2149.

15. Lee KW, Lee SJ, Kang NJ, Lee CY, & Lee HJ (2004). Effects of phenolics in Empire apples on hydrogen peroxide-induced inhibition of gap-junctional intercellular communication. *Biofactors* **21**, 361-365.

16. Li Y, Fu Z, Chen X, & Han R (1996). [Effects of curcumin derivatives on the GJIC of normal and tumor cells]. *Zhongguo Yi Xue Ke Xue Yuan Xue Bao* **18**, 111-115.

17. Livny O, Kaplan I, Reifen R, Polak-Charcon S, Madar Z, & Schwartz B (2002). Lycopene inhibits proliferation and enhances gap-junction communication of KB-1 human oral tumor cells. *J Nutr* **132**, 3754-3759.

18. Morita I, Zhang YW, & Murota SI (2001). Eicosapentaenoic acid protects endothelial cell function injured by hypoxia/reoxygenation. *Ann N Y Acad Sci* **947**, 394-397.

19. Na HK, Wilson MR, Kang KS, Chang CC, Grunberger D, & Trosko JE (2000). Restoration of gap junctional intercellular communication by caffeic acid phenethyl ester (CAPE) in a ras-transformed rat liver epithelial cell line. *Cancer Lett* **157**, 31-38.

20. Nakamura Y, Chang CC, Mori T, Sato K, Ohtsuki K, Upham BL, & Trosko JE (2005). Augmentation of differentiation and gap junction function by kaempferol in partially differentiated colon cancer cells. *Carcinogenesis* **26**, 665-671.

21. Nielsen M, Ruch RJ, & Vang O (2000). Resveratrol reverses tumor-promoter-induced inhibition of gap-junctional intercellular communication. *Biochem Biophys Res Commun* **275**, 804-809.

22. Park JU & Tsuchiya T (2002). Increase in gap junctional intercellular communication by high molecular weight hyaluronic acid associated with fibroblast growth factor 2 and keratinocyte growth factor production in normal human dermal fibroblasts. *Tissue Eng* **8**, 419-427.

23. Pelletier DB & Boynton AL (1994). Dissociation of PDGF receptor tyrosine kinase activity from PDGF-mediated inhibition of gap junctional communication. *J Cell Physiol* **158**, 427-434.

24. Rao AV & Agarwal S (2000). Role of antioxidant lycopene in cancer and heart disease. *J Am Coll Nutr* **19**, 563-569.

25. Remer T & Manz F (1995). Potential renal acid load of foods and its influence on urine pH. *J Am Diet Assoc* **95**, 791-797.

26. Sebastian A, Frassetto LA, Sellmeyer DE, Merriam RL, & Morris RC, Jr. (2002). Estimation of the net acid load of the diet of ancestral preagricultural Homo sapiens and their hominid ancestors. *Am J Clin Nutr* **76**, 1308-1316.

27. Sharov VS, Briviba K, & Sies H (1999). Peroxynitrite diminishes gap junctional communication: protection by selenite supplementation. *IUBMB Life* **48**, 379-384.

28. Sies H & Stahl W (1997). Carotenoids and intercellular communication via gap junctions. *Int J Vitam Nutr Res* **67**, 364-367.

29. Sleeman JM & Mudakikwa A.B. (1998). Analysis of Urine from Free-Range Mountain Gorillas for Normal Physiollogic Values. *Journal of Zoo and Wildlife Medicine* **29**, 432-434.

30. Spray DC, Fujita M, Saez JC, Choi H, Watanabe T, Hertzberg E, Rosenberg LC, & Reid LM (1987). Proteoglycans and glycosaminoglycans induce gap junction synthesis and function in primary liver cultures. *J Cell Biol* **105**, 541-551.

31. Stahl W & Sies H (1998). The role of carotenoids and retinoids in gap junctional communication. *Int J Vitam Nutr Res* **68**, 354-359.

32. Stahl W, von LJ, Martin HD, Emmerich T, & Sies H (2000). Stimulation of gap junctional communication: comparison of acyclo-retinoic acid and lycopene. *Arch Biochem Biophys* **373**, 271-274.

33. Stahl W, le-Agha N, & Polidori MC (2002). Non-antioxidant properties of carotenoids. *Biol Chem* **383**, 553-558.

34. Stoner GD & Mukhtar H (1995). Polyphenols as cancer chemopreventive agents. *J Cell Biochem Suppl* **22**, 169-180.

35. Takahashi C, Kikuchi N, Katou N, Miki T, Yanagida F, & Umeda M (1995). Possible anti-tumour-promoting activity of components in Japanese soybean fermented food, Natto: effect on gap junctional intercellular communication. *Carcinogenesis* **16**, 471-476.

36. Takahashi H, Nomata K, Mori K, Matsuo M, Miyaguchi T, Noguchi M, & Kanetake H (2004). The preventive effect of green tea on the gap junction intercellular communication in renal epithelial cells treated with a renal carcinogen. *Anticancer Res* **24**, 3757-3762.

37. Teicher VB, Kucharski N, Martin HD, van der SP, Sies H, & Stahl W (1999). Biological activities of Apo-canthaxanthinoic acids related to gap junctional communication. *Arch Biochem Biophys* **365**, 150-155.

38. Vine AL, Leung YM, & Bertram JS (2005). Transcriptional regulation of connexin 43 expression by retinoids and carotenoids: similarities and differences. *Mol Carcinog* **43**, 75-85.

39. Wolf G (1992). Retinoids and carotenoids as inhibitors of carcinogenesis and inducers of cell-cell communication. *Nutr Rev* **50**, 270-274.

40. Zhang LX, Cooney RV, & Bertram JS (1992). Carotenoids up-regulate connexin43 gene expression independent of their provitamin A or antioxidant properties. *Cancer Res* **52**, 5707-5712.

41. Zhang YW, Morita I, Yao XS, & Murota S (1999). Pretreatment with eicosapentaenoic acid prevented hypoxia/reoxygenation-induced abnormality in endothelial gap junctional intercellular communication through inhibiting the tyrosine kinase activity. *Prostaglandins Leukot Essent Fatty Acids* **61**, 33-40.

42. Zhang YW, Morita I, Nishida M, & Murota SI (1999). Involvement of tyrosine kinase in the hypoxia/reoxygenation-induced gap junctional intercellular communication abnormality in cultured human umbilical vein endothelial cells. *J Cell Physiol* **180** , 305-313.

43. Zhang YW, Dou DQ, Zhang L, Chen YJ, & Yao XS (2001). Effects of ginsenosides from Panax ginseng on cell-to-cell communication function mediated by gap junctions. *Planta Med* **67**, 417-422

Chapter 6 — Connexins and the mind-body connection

Exercise

1. Byrne HK & Wilmore JH (2001). The effects of a 20-week exercise training program on resting metabolic rate in previously sedentary, moderately obese women. *Int J Sport Nutr Exerc Metab* **11**, 15-31.

2. de Vries WR, Bernards NT, de Rooij MH, & Koppeschaar HP (2000). Dynamic exercise discloses different time-related responses in stress hormones. *Psychosom Med* **62**, 866-872.

3. Dipietro L, Williamson DF, Caspersen CJ, & Eaker E (1993). The descriptive epidemiology of selected physical activities and body weight among adults trying to lose weight: the Behavioral Risk Factor Surveillance System survey, 1989. *Int J Obes Relat Metab Disord* **17**, 69-76.

4. Dipietro L (1999). Physical activity in the prevention of obesity: current evidence and research issues. *Med Sci Sports Exerc* **31**, S542-S546.

5. Dipietro L (1999). Physical activity in the prevention of obesity: current evidence and research issues. *Med Sci Sports Exerc* **31**, S542-S546.

6. Donnelly JE, Pronk NP, Jacobsen DJ, Pronk SJ, & Jakicic JM (1991). Effects of a very-low-calorie diet and physical-training regimens on body composition and resting metabolic rate in obese females. *Am J Clin Nutr* **54**, 56-61.

7. Drewnowski A & Evans WJ (2001). Nutrition, physical activity, and quality of life in older adults: summary. *J Gerontol A Biol Sci Med Sci* **56 Spec No 2**, 89-94.

8. Frandin K, Mellstrom D, Sundh V, & Grimby G (1995). A life span perspective on patterns of physical activity and functional performance at the age of 76. *Gerontology* **41**, 109-120.

9. Grodstein F, Levine R, Troy L, Spencer T, Colditz GA, & Stampfer MJ (1996). Three-year follow-up of participants in a commercial weight loss program. Can you keep it off? *Arch Intern Med* **156**, 1302-1306.

10. Hendley O.F. (1939). Analysis of 5116 deaths reported as due to acute coronary occlusion in Philadelphia, 1933-1936. *US Weekly Public Health Report* **54**, 972ff.

11. Herrick JB (1912). Clinical features of sudden obstruction of the coronary arteries. *JAMA* **59**, 2015-2020.

12. Lee IM, Hsieh CC, & Paffenbarger RS, Jr. (1995). Exercise intensity and longevity in men. The Harvard Alumni Health Study. *JAMA* **273**, 1179-1184.

13. Lee IM, Paffenbarger RS, Jr., & Hennekens CH (1997). Physical activity, physical fitness and longevity. *Aging (Milano)* **9**, 2-11.

14. Lee IM & Paffenbarger RS, Jr. (2000). Associations of light, moderate, and vigorous intensity physical activity with longevity. The Harvard Alumni Health Study. *Am J Epidemiol* **151**, 293-299.

15. Lee IM, Sesso HD, Oguma Y, & Paffenbarger RS, Jr. (2003). Relative intensity of physical activity and risk of coronary heart disease. *Circulation* **107**, 1110-1116.

16. Lee IM, Sesso HD, Oguma Y, & Paffenbarger RS, Jr. (2004). The "weekend warrior" and risk of mortality. *Am J Epidemiol* **160**, 636-641.

17. Lissner L, Bengtsson C, Bjorkelund C, & Wedel H (1996). Physical activity levels and changes in relation to longevity. A prospective study of Swedish women. *Am J Epidemiol* **143**, 54-62.

18. Oguma Y, Sesso HD, Paffenbarger RS, Jr., & Lee IM (2002). Physical activity and all cause mortality in women: a review of the evidence. *Br J Sports Med* **36**, 162-172.

19. Paffenbarger RS, Jr., Blair SN, & Lee IM (2001). A history of physical activity, cardiovascular health and longevity: the scientific contributions of Jeremy N Morris, DSc, DPH, FRCP. *Int J Epidemiol* **30**, 1184-1192.

20. Ramazzini B (1940). *De Morbis Artificum Diatriba (Translated fromn the Latin text of 1713, revised with translation and notes by Wilmer Cave Wright)* The University of Chicago Press, Chicago.

21. Sawada SS, Muto T, Tanaka H, Lee IM, Paffenbarger RS, Jr., Shindo M, & Blair SN (2003). Cardiorespiratory fitness and cancer mortality in Japanese men: a prospective study. *Med Sci Sports Exerc* **35**, 1546-1550.

22. Seguin R & Nelson ME (2003). The benefits of strength training for older adults. *Am J Prev Med* **25**, 141-149.

23. Sesso HD, Paffenbarger RS, Jr., & Lee IM (2000). Physical activity and coronary heart disease in men: The Harvard Alumni Health Study. *Circulation* **102**, 975-980.

24. Sherman SE, D'Agostino RB, Cobb JL, & Kannel WB (1994). Does exercise reduce mortality rates in the elderly? Experience from the Framingham Heart Study. *Am Heart J* **128**, 965-972.

25. Stessman J, Hammerman-Rozenberg R, Maaravi Y, & Cohen A (2002). Effect of exercise on ease in performing activities of daily living and instrumental activities of daily living from age 70 to 77: the Jerusalem longitudinal study. *J Am Geriatr Soc* **50**, 1934-1938.

26. Stessman J, Hammerman-Rozenberg R, Maaravi Y, Azoulai D, & Cohen A (2005). Strategies to enhance longevity and independent function: the Jerusalem Longitudinal Study. *Mech Ageing Dev* **126**, 327-331.

27. Sundberg CJ & Jansson E (1998). [Reduced morbidity and the risk of premature death. Regular physical exercise is beneficial for health at all ages]. *Lakartidningen* **95**, 4062-4067.

28. Thornton EW, Sykes KS, & Tang WK (2004). Health benefits of Tai Chi exercise: improved balance and blood pressure in middle-aged women. *Health Promot Int* **19**, 33-38.

Mind-Body: Stress

1. Blomstrand F, Giaume C, Hansson E, & Ronnback L (1999). Distinct pharmacological properties of ET-1 and ET-3 on astroglial gap junctions and Ca(2+) signaling. *Am J Physiol* **277**, C616-C627.

2. Blomstrand F, Venance L, Siren AL, Ezan P, Hanse E, Glowinski J, Ehrenreich H, & Giaume C (2004). Endothelins regulate astrocyte gap junctions in rat hippocampal slices. *Eur J Neurosci* **19**, 1005-1015.

3. Bonnet F, Irving K, Terra JL, Nony P, Berthezene F, & Moulin P (2005). Anxiety and depression are associated with unhealthy lifestyle in patients at risk of cardiovascular disease. *Atherosclerosis* **178**, 339-344.

4. Callaghan P (2004). Exercise: a neglected intervention in mental health care? *J Psychiatr Ment Health Nurs* **11**, 476-483.

5. Castellani S, Ungar A, La CG, Cantini C, Stefanile C, Camaiti A, Messeri G, Coppo M, Vallotti B, Di SC, Brocchi A, & Masotti G (1997). Renal adaptation to stress: a possible role of endothelin release and prostaglandin modulation in the human subject. *J Lab Clin Med* **129**, 462-469.

6. Cramer SR, Nieman DC, & Lee JW (1991). The effects of moderate exercise training on psychological well-being and mood state in women. *J Psychosom Res* **35**, 437-449.

7. Crews DJ, Lochbaum MR, & Landers DM (2004). Aerobic physical activity effects on psychological well-being in low-income Hispanic children. *Percept Mot Skills* **98**, 319-324.

8. De Mello WC (1996). Renin-angiotensin system and cell communication in the failing heart. *Hypertension* **27**, 1267-1272.

9. Giepmans BN, Hengeveld T, Postma FR, & Moolenaar WH (2001). Interaction of c-Src with gap junction protein connexin-43. Role in the regulation of cell-cell communication. *J Biol Chem* **276**, 8544-8549.

10. Giepmans BN, Feiken E, Gebbink MF, & Moolenaar WH (2003). Association of connexin43 with a receptor protein tyrosine phosphatase. *Cell Commun Adhes* **10**, 201-205.

11. Goodwin RD (2003). Association between physical activity and mental disorders among adults in the United States. *Prev Med* **36**, 698-703.

12. Guszkowska M (2004). [Effects of exercise on anxiety, depression and mood]. *Psychiatr Pol* 38, 611-620.

13. Kawamura H, Kobayashi M, Li Q, Yamanishi S, Katsumura K, Minami M, Wu DM, & Puro DG (2004). Effects of angiotensin II on the pericyte-containing microvasculature of the rat retina. *J Physiol* **561**, 671-683.

14. Knechtle B (2004). [Influence of physical activity on mental well-being and psychiatric disorders]. *Schweiz Rundsch Med Prax* **93**, 1403-1411.

15. Lavie CJ & Milani RV (2004). Prevalence of anxiety in coronary patients with improvement following cardiac rehabilitation and exercise training. *Am J Cardiol* **93**, 336-339.

16. Manger TA & Motta RW (2005). The impact of an exercise program on posttraumatic stress disorder, anxiety, and depression. *Int J Emerg Ment Health* 7, 49-57.

17. Mangiafico RA, Malatino LS, Attina T, Messina R, & Fiore CE (2002). Exaggerated endothelin release in response to acute mental stress in patients with intermittent claudication. *Angiology* **53**, 383-390.

18. Netz Y & Lidor R (2003). Mood alterations in mindful versus aerobic exercise modes. *J Psychol* **137**, 405-419.

19. North TC, McCullagh P, & Tran ZV (1990). Effect of exercise on depression. *Exerc Sport Sci Rev* **18**, 379-415.

20. Osei-Tutu KB & Campagna PD (2005). The effects of short- vs. long-bout exercise on mood, VO2max, and percent body fat. *Prev Med* **40**, 92-98.

21. Petruzzello SJ, Landers DM, Hatfield BD, Kubitz KA, & Salazar W (1991). A meta-analysis on the anxiety-reducing effects of acute and chronic exercise. Outcomes and mechanisms. *Sports Med* **11**, 143-182.

22. Polontchouk L, Ebelt B, Jackels M, & Dhein S (2002). Chronic effects of endothelin 1 and angiotensin II on gap junctions and intercellular communication in cardiac cells. *FASEB J* **16**, 87-89.

23. Rozyczka J, Figiel M, & Engele J (2005). Chronic endothelin exposure inhibits connexin43 expression in cultured cortical astroglia. *J Neurosci Res* **79**, 303-309.

24. Schmitz N, Kruse J, & Kugler J (2004). The association between physical exercises and health-related quality of life in subjects with mental disorders: results from a cross-sectional survey. *Prev Med* **39**, 1200-1207.

25. Shiokawa-Sawada M, Mano H, Hanada K, Kakudo S, Kameda T, Miyazawa K, Nakamaru Y, Yuasa T, Mori Y, Kumegawa M, & Hakeda Y (1997). Down-regulation of gap junctional intercellular communication between osteoblastic MC3T3-E1 cells by basic fibroblast growth factor and a phorbol ester (12-O-tetradecanoylphorbol-13-acetate). *J Bone Miner Res* **12**, 1165-1173.

26. Spinella F, Rosano L, Di C, V, Nicotra MR, Natali PG, & Bagnato A (2003). Endothelin-1 decreases gap junctional intercellular communication by inducing phosphorylation of connexin 43 in human ovarian carcinoma cells. *J Biol Chem* **278**, 41294-41301.

27. Treiber FA, Kapuku GK, Davis H, Pollock JS, & Pollock DM (2002). Plasma endothelin-1 release during acute stress: role of ethnicity and sex. *Psychosom Med* **64**, 707-713.

28. Ueyama T, Aki T, Senba E, Hatanaka K, & Yoshida K (2001). Emotional stress activates MAP kinase in the rat heart. *Life Sci* **69**, 1927-1934.

29. Zhou L, Kasperek EM, & Nicholson BJ (1999). Dissection of the molecular basis of pp60(v-src) induced gating of connexin 43 gap junction channels. *J Cell Biol* **144**, 1033-1045.

Chapter 7 — So let's get started

1. Remer T & Manz F (1995). Potential renal acid load of foods and its influence on urine pH. *J Am Diet Assoc* **95**, 791-797.

Chapter 11 — Who's monitoring those connexins?

1. Agarwal R (1999). Role of home blood pressure monitoring in hemodialysis patients. *Am J Kidney Dis* **33**, 682-687.

2. Cappuccio FP, Kerry SM, Forbes L, & Donald A (2004). Blood pressure control by home monitoring: meta-analysis of randomised trials. *BMJ* **329**, 145.

3. Den HE, Staessen JA, Celis H, Fagard R, Keary L, Vandenhoven G, & O'Brien ET (2004). Antihypertensive treatment based on home or office blood pressure--the THOP trial. *Blood Press Monit* **9**, 311-314.

4. Ellington TM & Stolte S (2000). Patient assessment tools: utilizing diagnostic and monitoring devices. *J Am Pharm Assoc (Wash)* **40**, S58-S59.

5. Goldberg HI, Ralston JD, Hirsch IB, Hoath JI, & Ahmed KI (2003). Using an Internet comanagement module to improve the quality of chronic disease care. *Jt Comm J Qual Saf* **29**, 443-451.

6. Hozawa A, Ohkubo T, Kikuya M, Ugajin T, Yamaguchi J, Asayama K, Metoki H, Ohmori K, Hoshi H, Hashimoto J, Satoh H, Tsuji I, & Imai Y (2004). Prognostic value of home heart rate for cardiovascular mortality in the general population: the Ohasama study. *Am J Hypertens* **17**, 1005-1010.

7. Imai Y, Ohkubo T, Sakuma M, Tsuji I, I, Satoh H, Nagai K, Hisamichi S, & Abe K (1996). Predictive power of screening blood pressure, ambulatory blood pressure and blood pressure measured at home for overall and cardiovascular mortality: a prospective observation in a cohort from Ohasama, northern Japan. *Blood Press Monit* **1**, 251-254.

8. Imai Y, Otsuka K, Kawano Y, Shimada K, Hayashi H, Tochikubo O, Miyakawa M, & Fukiyama K (2003). Japanese society of hypertension (JSH) guidelines for self-monitoring of blood pressure at home. *Hypertens Res* **26**, 771-782.

9. Inoue R, Ohkubo T, Kikuya M, & Imai Y (2005). [A clinical significance of home blood pressure measurements for diagnosis of hypertension in the elderly]. *Nippon Rinsho* **63**, 1005-1009.

10. Kamoi K, Miyakoshi M, Soda S, Kaneko S, & Nakagawa O (2002). Usefulness of home blood pressure measurement in the morning in type 2 diabetic patients. *Diabetes Care* **25**, 2218-2223.

11. Kamoi K, Imamura Y, Miyakoshi M, & Kobayashi C (2003). Usefulness of home blood pressure measurement in the morning in type 1 diabetic patients. *Diabetes Care* **26**, 2473-2475.

12. Mallion JM, Genes N, Vaur L, Clerson P, Vaisse B, Bobrie G, & Chatellier G (2004). Detection of masked hypertension by home blood pressure measurement: is the number of measurements an important issue? *Blood Press Monit* **9**, 301-305.

13. Mansoor GA & White WB (2004). Self-measured home blood pressure in predicting ambulatory hypertension. *Am J Hypertens* **17**, 1017-1022.

14. Miyakawa M (2005). [Clinical significance of home blood pressure monitoring and its application to daily clinical practice]. *Nippon Rinsho* **63**, 909-915.

15. Ohkubo T, Asayama K, Kikuya M, Metoki H, Obara T, Saito S, Hoshi H, Hashimoto J, Totsune K, Satoh H, & Imai Y (2004). Prediction of ischaemic and haemorrhagic stroke by self-measured blood pressure at home: the Ohasama study. *Blood Press Monit* **9**, 315-320.

16. Penny JA, Halligan AW, Shennan AH, Lambert PC, Jones DR, De SM, & Taylor DJ (1998). Automated, ambulatory, or conventional blood pressure measurement in pregnancy: which is the better predictor of severe hypertension? *Am J Obstet Gynecol* **178**, 521-526.

17. Rickerby J (2002). The role of home blood pressure measurement in managing hypertension: an evidence-based review. *J Hum Hypertens* **16**, 469-472.

18. Rickerby J & Woodward J (2003). Patients' experiences and opinions of home blood pressure measurement. *J Hum Hypertens* **17**, 495-503.

19. Saito I, I, Takeshita E, Murata K, Kawabe H, & Saruta T (1996). Serum cortisol in the white-coat phenomenon. *Blood Press Monit* **1**, 381-383.

20. Sakuma M, Imai Y, Nagai K, Watanabe N, Sakuma H, Minami N, Satoh H, & Abe K (1997). Reproducibility of home blood pressure measurements over a 1-year period. *Am J Hypertens* **10**, 798-803.

21. Tachibana R, Tabara Y, Kondo I, Miki T, & Kohara K (2004). Home blood pressure is a better predictor of carotid atherosclerosis than office blood pressure in community-dwelling subjects. *Hypertens Res* **27**, 633-639.

22. Vaur L, Dubroca I, I, Dutrey-Dupagne C, Genes N, Chatellier G, Bouvier-d'Yvoire M, Elkik F, & Menard J (1998). Superiority of home blood pressure measurements over office measurements for testing antihypertensive drugs. *Blood Press Monit* **3**, 107-114.

23. Waugh J, Bosio P, Habiba M, Boyce T, Shennan A, & Halligan A (2001). Home monitoring of blood pressure in pregnancy at high risk of pre-eclampsia. *Eur J Obstet Gynecol Reprod Biol* **99**, 109-111.

24. Waugh J, Habiba MA, Bosio P, Boyce T, Shennan A, & Halligan AW (2003). Patient initiated home blood pressure recordings are accurate in hypertensive pregnant women. *Hypertens Pregnancy* **22**, 93-97.

25. Weisser B, Mengden T, Dusing R, Vetter H, & Vetter W (2000). Normal values of blood pressure self-measurement in view of the 1999 World Health Organization-International Society of Hypertension guidelines. *Am J Hypertens* **13**, 940-943.

ABOUT THE AUTHOR

A life of science. A life of medicine.
A life of nutrition, health and wellness.

Dr. Tanelian began life as a child fascinated by the workings of the universe. At age 15 he began his formal scientific education conducting graduate level research into the neurophysiology of salmon homing conducted by the University of Wisconsin. At the same time he independently reversed radiation-induced illness and death in animals by creating an injectable thymus polypeptide fraction for which he was awarded the Bausch and Lomb Science Medal.

Encouraged by this success, he went on to Stanford University where he obtained a B.S., M.D., Ph.D., internship, residency and fellowship all in preparation for a career in academic medicine.

Subsequently, he realized this life-long dream and held professorships and faculty positions in Neurology, Neuroscience, Engineering, Anesthesiology and Pain Management at Stanford University School of Medicine and The University of Texas, Southwestern Medical School. Dr. Tanelian's research received one of the highest priority scores at the National Institutes of Health (NIH) which awarded him grant funding, along with numerous other granting agencies. His research culminated in over 100 peer-reviewed scientific and medical publications, as well as in a biophysics book describing the fundamentals of life, *Liquid Interfaces in Chemistry and Biology.* At U.T Southwestern, he was honored by holding the Jane and Bill Browning, Jr., Chair in Medical Science.

Now, Dr. Tanelian has reformulated the basic foundations of science and medicine into a molecular health and wellness program – *The Connexin Connection Molecular Medicine Program.* Along the way, he obtained a M.B.A from Southern Methodist University and business experience in the biotechnology and pharmaceutical industries to best compliment his scientific and medical expertise, preparing him to bring this program to you. He gives the world the Connexin Molecular Fitness Lifestyle in hopes of transforming your health and wellbeing and that of future generations to come.